Chemical Bodies

Geopolitical Bodies, Material Worlds

This series publishes studies that originate in a range of different fields that are nonetheless linked through their common foundation: a belief that the macro-scale of geopolitics is composed of trans-local relations between bodies and materials that are only understandable through empirical examination of those relations. It is the interaction of these elements that produces the forces that shape global politics, often with outcomes that differ from the predictions of macro-scaled theories. This world poses questions: How do materialities such as the built environment and the body reproduce global power structures? How are they caught up in violent transformations and how do they become sites of resistance? How do assemblages of human and non-human elements both fortify and transform political space? What possibilities for political change are latent within the present?

Series Editors:

Jason Dittmer, Professor of Political Geography at University College London

Ian Klinke, Associate Professor in Human Geography and Tutorial Fellow at St John's College, Oxford

Titles in the Series

The Geopolitics of Real Estate: Reconfiguring Property, Capital and Rights, Dallas Rogers

Choreographies of Resistance: Mobilities, Bodies and Politics, Tarja Väyrynen, Eeva Puumala, Samu Pehkonen, Tiina Vaittinen and Anitta Kynsilehto

Life Adrift: Climate Change, Migration, Critique, Edited by Andrew Baldwin and Giovanni Bettini

Territory beyond Terra, Edited by Kimberley Peters, Philip Steinberg and Elaine Stratford

Chemical Bodies: The Techno-Politics of Control, Edited by Alex Mankoo and Brian Rappert

Forthcoming:

Camps Revisited: Multifaceted Spatialities of a Modern Political Technology, Edited by Irit Katz, Diana Martin and Claudio Minca

Micropolitics on the Line: Excavating Earth and Nature in Ecuador, Joe Gerlach

Submarine Geopolitics: Living under the Sea through the Sealab Experiments, Rachael Squire

Chemical Bodies

The Techno-Politics of Control

Edited by Alex Mankoo and
Brian Rappert

ROWMAN &
LITTLEFIELD
INTERNATIONAL
London • New York

Published by Rowman & Littlefield International Ltd.
6 Tinworth Street, London SE11 5AL, United Kingdom
www.rowmaninternational.com

Rowman & Littlefield International Ltd. is an affiliate of Rowman & Littlefield
4501 Forbes Boulevard, Suite 200, Lanham, Maryland 20706, USA
With additional offices in Boulder, New York, Toronto (Canada) and Plymouth (UK)
www.rowman.com

British Library Cataloguing in Publication Data

A catalogue record for this book is available from the British Library

ISBN: HB 978-1-7866-0586-3

Library of Congress Cataloging-in-Publication Data Is Available

978-1-78660-586-3 (cloth: alk. paper)
978-1-78660-587-0 (electronic)

Contents

Acknowledgements

The idea for this volume came together following a conference at University College London (UCL) titled 'Science/Technology/Security: Challenges to Global Governance?'. The conference was organized by Jason Dittmer and Alex Mankoo and hosted through UCL's Global Governance Institute; most of the contributors here were chosen following presentations they gave at this event. Thanks to those who attended and contributed to discussions at the conference. The editors thank the series editors, Jason Dittmer and Ian Klinke; Natalie Linh Bolderston, Dhara Snowden and Rebecca Anastasi at Rowman and Littlefield for their editorial assistance; and all the contributors.

In addition, Alex Mankoo would like to thank the Economic and Social Research Council (ESRC) for funding his PhD research at UCL (grant ES/J500185/1), which ultimately led to the opportunity to put this volume together. He would also like to thank his wife Dianna, his supervisors, Brian Balmer and John Agar, and colleagues in both the UCL STS department and Harvard Kennedy School STS programme for their support.

With acknowledgements for permission to include copyrighted material to TRACE for the *Floorplan for Poisoned Pasts* and to Nelson Mandela Foundation for the photo title 'Affective Panels'. Elements of the chapter 'Opening Spaces through Exhibiting Absences: Representing Secretive Pasts' have been drawn from the book *Dis-eases of Secrecy: Tracing History, Memory and Justice* by Brian Rappert and Chandré Gould, with permission from Jacana Media. Anna Feigenbaum's chapter 'Tear Gas and Colonial Bodies in the British Interwar Period' has been adapted from her recent book *Tear Gas: From the Battlefields of World War I to the Streets of Today* (Verso, 2017).

Abbreviations

AHRC	Arts and Humanities Research Council
ANC	African National Congress
ARP	Air Raid Precautions
BBC	British Broadcasting Corporation
BSSRS	British Society for Social Responsibility in Science
BTWC	Biological and Toxin Weapons Convention [also the BWC]
BW	Biological warfare/weapons
BWC	Biological Weapons Convention (also the BTWC)
CAP	Chloroacetophenone (see also CN)
CBDE	Chemical and Biological Defence Establishment
CBM	Confidence building measures
CBW	Chemical and biological weapons/warfare
CDAB	Chemical Defence Advisory Board
CDEE	Chemical Defence Experimental Establishment
CDES	Chemical Defence Experimental Station
CIA	Central Intelligence Agency (US)
CN	Chloroacetophenone
CNS	Central nervous system
CR	Dibenzoxazepine
CS	2-chlorobenzalmalononitrile
CSP	Conference of States Parties
CW	Chemical warfare/weapons
CWC	Chemical Weapons Convention
CWS	Chemical Warfare Service (US)
DDT	Dichlorodiphenyltrichloroethane
DOD	Department of Defense (US)
DSTL	Defence Science and Technology Laboratory (UK)

ENMOD	Convention on the Prohibition of Military or Any Other Hostile Use of Environmental Modification Techniques
ESRC	Economic and Social Research Council
FCO	Foreign and Commonwealth Office (UK)
GP	Geneva Protocol
GPC	General purpose criterion
HPCSA	Health Professionals Council of South Africa
ICA	Incapacitating chemical agent
ICCPR	International Covenant on Civil and Political Rights
ICRC	International Committee of the Red Cross
IMT	International Military Tribunal
IMTFE	International Military Tribunal for the Far East
IPS	International Prosecution Section
LS	(Allied) Legal Section
MoD	Ministry of Defence (UK)
MRC	Medical Research Council (UK)
NGOs	Non-governmental organizations
NLW	Non-lethal weapon
OEWG	Open-ended working group
OPCW	Organisation for the Prohibition of Chemical Weapons
POW	Prisoner of war
PR	Public relations
RCA	Riot control agent
RUC	Royal Ulster Constabulary
SADF	South African Defence Force
SCAP	Supreme Commander for the Allied Powers
SMS	Short message service
STS	Science and technology studies
SWAPO	South West African People's Organization
TNA	The National Archives (UK)
TRC	Truth and Reconciliation Commission
UNDHR	Universal Declaration on Human Rights
WMD	Weapons of Mass Destruction

Chapter 1

Transgressive Chemicals

Brian Rappert and Alex Mankoo

When President Donald Trump spoke on the evening of 6 April 2017 regarding his decision to bomb a Syrian airfield, his brief statement raised many themes commonly associated with chemical weapons (CW) today. He began:

> My fellow Americans, on Tuesday, Syrian dictator Bashar al-Assad launched a horrible chemical weapons attack on innocent civilians using a deadly nerve agent. Assad choked out the lives of helpless men, women, and children. It was a slow and brutal death for so many – even beautiful babies were cruelly murdered in this very barbaric attack. No child of God should ever suffer such horror.[1]

With these words, definite markers were given for what happened and how it should be understood: a user was named (Bashar al-Assad), characterized (as a 'dictator') and deemed directly responsible ('Assad choked out the lives'). The weaponry involved was classified (a 'deadly nerve agent'), its severe effects depicted (choking, 'a slow and brutal death') and the victimhood standing of the groups targeted forwarded ('innocent civilians', 'beautiful babies' and 'helpless men, women and children').

Just as the condemnation given typified the moral charge so often associated with chemical weapons, the intense international debate about the Trump Administration's action epitomized the manner in which violence can be contested. So it was asked: Were the images shared around the world of people dying genuine? Were the attacks carried out by the Syrian military under instruction from the president? Was it really the images of beautiful babies suffering that led Trump to strike or were ulterior geopolitical, domestic or personal motivations in play? Since many beautiful babies had died as part of the long-running Syrian conflict, why were these singled out as justifying a U.S. military response? Would this attack serve to escalate violence or help turn warring factions towards some kind of resolution? If the Syrian men, women and children were so blameless, why was the Trump administration simultaneously seeking an outright ban on Syrian refugees entering the United States?[2]

While this was a highly prominent instance of the hostile use of chemicals, other events also took place around the same time. In the early hours of 17 April 2017, twenty people were injured at a nightclub in London in what was widely described as an 'acid attack'. Reports indicated a chemical was sprayed as part of a fight.[3] Among those assaulted, two were blinded in one eye. One woman affected commented, 'The pain is indescribable – it feels like your skin is eating itself. The police told us later it was drain fluid'.[4] The attack was notable for the extent of media coverage given to it compared to previous acid attacks in the United Kingdom – the latter often undertaken against women by their ex-partners in displays of power.[5] A recurring feature of the coverage in April was that one of the three men suspected as carrying out the assault was the former boyfriend of a British reality TV star. While the nightclub attack was given extraordinary media attention, this instance would be followed by prominent coverage to a number of similar attacks in public. One reason identified for the rise in incidents was the clampdown on other means of force, notably knives.[6] Through this cumulative attention, 'acid attacks' would move from the fringe to the centre of criminal justice policy in Britain.

The attention to the previous two cases dwarfed that given to another set of events taking place around the same time in Western media. As reported by Amnesty International, the days before the sixth anniversary of the 2011 uprising in Bahrain (17 February 2017) amounted to a human rights crisis. In describing the violence employed as part of the crackdown on the freedom of expression, Amnesty International repeatedly cited the use of tear gas against protesters.[7] For instance, in relation to large-scale protests in the villages of Duraz, Sitra and al-Daih, 'security forces were said to have responded with tear gas and shotguns firing birdshot. Eyewitnesses told Amnesty International that they saw security officers in Sanabis fire tear gas, aiming directly at protesters and causing injuries'.[8]

These three contrasting cases exemplify the array of co-existing and shifting notions of what chemical agents are, as well as assessment of how, when and by whom they should (or should not) be used.

Chemical Bodies takes as its topic how chemical agents have been justified and resisted as instruments of coercion across time and space. In taking the potential for contestation as a central concern, this volume explores how the use of chemical agents often entails messy assemblages of materials, equipment, individuals, techniques, modes of argument and forms of accountability. In seeking to understand how chemical agents are made to matter, the contributors to this volume examine how certain instances of force gain prominence (or not), how some individuals speak and others get spoken for, how definitions of what counts as 'success' and 'failure' are advanced and how the normative rights and wrongs of violence are justified. *Chemical Bodies* aims to navigate the relationship between chemicals, bodies and space to answer questions such as the following:

- How, when and for whom do bodies become rendered as sites of intervention?
- What kind of geopolitical spaces make body-controlling chemicals possible and permissible?
- What concerns and whose experiences are made visible and invisible?
- How can the analysis of the past inform scholarship, policy making and activism?

This introductory chapter provides an overview of how these weapons have been shaped by scientific and technical advancements, geopolitics, international law and normative standards. These matters will be explored through an analysis that seeks to develop sensitivities for approaching disputes about the use of chemicals; sensitivities opened up to in subsequent chapters. The next section considers how distinctions have been drawn over time in defining and assessing 'chemical weapons'. Determining whether chemical agents are unlawful, unadvisable, offensive, prohibited or reprehensible entails distinguishing them from what is legal, advisable, defensive, allowed or above board. Building on this, the third section analyses how attempts to legitimate and contest chemical weapons have entailed efforts to link and delink contexts, technologies and forms of expertise from each other. In the manner that chemical agents have been embraced as a means for sorting individuals and spaces as well as resisted as unleashing forms of uncontrollable contamination, their introduction has entailed experimentation with bodies and spaces. The fourth section examines how the notion of experimentation has figured as part of contests about chemical force.

DISTINCTIONS

Distinctions matter. They cordon off parts of the world from each other in order to distinguish events, objects, locales and so on. Events, objects and so on must have their limits if the world is not to be treated as one indistinguishable goo. While distinctions are needed, too many would result in an unruly and confusing scattering of unique objects. In marking off certain acts or objects, distinctions bring into effect relations of resemblance and difference.

For instance, is the kind of 'acid attack' in London mentioned earlier properly put under the umbrella of a 'chemical weapon' attack? If the determination is 'yes', then a host of imaginations and concerns would be evoked associated with their terror potential or their status as an instance of 'weapons of mass destruction' (WMD). These concerns, however, were not prevalent in public coverage. In taking as its concern 'chemical agents', this volume seeks to bring into view a range of coercive uses of chemicals that might fall outside of analyses concerned with a narrow definition of 'chemical weapons'.

Distinctions matter as well in the sense of telling us what matters, what is important. This point is highly relevant to the topics at hand in these chapters because the use of 'chemical weapons' is often highly charged. As with terms like 'violence', to say someone has used chemical weapons is typically not taken as a mere description. It comes pre-loaded as an evaluative category. Aversion, disgust and repulsion are the types of emotional recoil often connected with chemical weapons, not least because of the manner they are associated with bodily contamination, pollution and defilement. Although to a lesser extent, much the same spark characterizes associated words. In news reports, terms like 'tear gas' regularly stand as bywords suggestive of severe disorder or iron-fist policing.[9]

While often charged, the meaning of categories formed from distinctions should not be taken as invariant or context-free. In 'From Reviled Poisons to State Arsenals: The Un(necessary) Proliferation of Chemical Weapons', Jeanne Guillemin sets the development and the post–World War II protection of Japan from prosecution for its chemical warfare (CW) in China against three distinct phases in the legal restraint of chemical weapons in the twentieth century; each which offered a particular set of prevalent meanings. As she argues, in the decade after the Great War, the moral revulsion for poison weapons supported universal legal restrictions; foremost among them was the 1925 Geneva Protocol (GP) that forbade their use in war. Beginning in the early 1930s, support for the universal rules of war declined in a new age of total war theory and the rise of the Axis powers, two of which with impunity conducted chemical warfare against defenceless enemies – Italy against Ethiopia and Japan against China. Guillemin's third phase began with the 1945 U.S. dropping of atomic bombs on Japan and the dawn of the nuclear age. As she maintains, the ill-considered 1948 United Nations (UN) definition of chemical weapons as 'WMD' accorded them strategic value.

James Revill and Marcos Favero's chapter covers much of the same time period as Guillemin, but they offer a more thematic take on the assessments made of the use of chemical weapons. 'Lesser Appreciations: A History of Interwar Chemical Warfare' takes as its primary focus how distinctions between those 'ingroup' and 'outgroup' as well as 'civilized' and 'other' has enabled otherwise-'unacceptable' weapons to be rendered acceptable or at least tolerable.

Moving from such wide-ranging historical analyses, the meaning of categories formed from distinctions is also accomplished as part of situated, day-to-day interactions.[10] Consider both the importance of distinctions and the potential scope for their negotiation in relation to the issue of what counts as a 'chemical weapon'. When Sean Spicer, then press secretary for the White House, stood before reporters on 11 April, high on the list of questions were ones related to the sarin attacks that took place a week earlier on the Syrian town of Khan Shaykhun and the Trump administration's subsequent airstrike on 6 April. In responding to questions from the floor, Spicer situated chemical weapons as types of 'WMD'. Shortly after doing so the following exchange took place:

> *Reporter*: Thanks a lot, Sean. The alliance between Russia and Syria is a strong one; it goes back decades. President Putin has supplied personnel; he's supplied military equipment to the Assad government. What makes you think that at this point, he's going to pull back in his support for President Assad and for the Syrian government right now?
>
> *Sean Spicer*: I think a couple things. You look – we didn't use chemical weapons in World War II. You had someone as despicable as Hitler who didn't even sink to using chemical weapons. So you have to, if you're Russia, ask yourself is this a country that you and a regime that you want to align yourself with? You have previously signed on to international agreements rightfully acknowledging that the use of chemical weapons should be out of bounds by every country. To not stand up to not only Assad, but your own word, should be troubling.
>
> Russia put their name on the line. So it's not a question of how long that alliance has lasted, but at what point do they recognize that they are now getting on the wrong side of history in a really bad way really quickly.[11]

As part of this exchange, when asked a question about why Russia would stop its long-running support for the Assad government in Syria, Spicer initially responds by mobilizing a negative evaluation of chemical weapons. He does so by offering a couple contrasts, one with the non-use of these weapons by the United States in World War II and the second the non-use of chemical weapons by Hitler in the same war. With regard to the latter, the comparison reference to Hitler makes use of a culturally widespread moral marker. Doing so not only brings together Hitler and Assad for the purpose of setting them apart from other world leaders, the Syrian president is positioned as even going beyond the taken-for-granted deplorable status of Hitler. This is so because the latter 'didn't even sink to using chemical weapons'.[12] In this way, the use of chemical weapons gets morally set apart from other kinds of violence.

The attempt to justify a distinct standing for Assad through linking him with CW would subsequently be queried later in the briefing:

> *Reporter 2*: Sean, thanks. I just want to give you an opportunity to clarify something you said that seems to be gaining some traction right now. 'Hitler didn't even sink to the level of using chemical weapons'. What did you mean by that?
>
> *Sean Spicer*: I think when you come to sarin gas, there was no – he was not using the gas on his own people the same way that Assad is doing, I mean, there was clearly – I understand your point, thank you.
>
> *Reporter 2*: I'm just getting –
>
> *Sean Spicer*: Thank you, I appreciate that. There was not – he brought them into the Holocaust centre, I understand that. But I'm saying in the way that Assad used them, where he went into towns, dropped them down to innocent – into the middle of towns. It was brought – so the use of it – I appreciate the clarification there. That was not the intent.

In her initial remarks, this reporter pitches her question in the form of a requirement for clarification, not a criticism; this face-value purpose is subsequently supported by the phrasing of her question in the open-ended, non-directed formulation of 'What did you mean by that?'. In reformulating his depiction of what Hitler did, Spicer offers the qualification that Hitler had not used chemical weapons 'the same way that Assad is doing' and acknowledges a need for clarification in relation to a point, but what that point is remains unstated and indefinite.

Not in the official White House transcript but reported elsewhere, after the initial response by Spicer and overlapping with the interjection by Reporter 2 of 'I'm just getting – ', another reporter present was said to shout off mike that Jews had been targeted.[13] Spicer's follow-on comments then speak to the plight of Jews under Nazi Germany with allusion to the gas chambers and further his previous qualification about the difference between Assad and Hitler by contending the former 'went into towns, dropped them down to innocent – into the middle of towns'. In other words, a distinction seems to be offered on the basis of how chemicals were used, rather than that they in and of themselves were used. While the remarks by Spicer might be taken as question begging in relation to what attributions of sameness and difference are being made, the referential adequacy of some claims and the meaning of terms such as 'Holocaust centre', at the time in the press briefing, no further clarifications were sought or given.

Subsequent media coverage though would make relevant a host of points. In relation to the question of whether chemical weapons were used, commentators would note that millions of Jews and others had been killed during the Holocaust, including through the use of 'chemical gas agents'.[14] The National Jewish Democratic Council would be quoted in one report as stating 'Hitler used chemical weapons. Period. No amount of clarification or walking back from the press secretary's office should be accepted for Spicer's horrific mistake. We demand a full apology and a promise that these sorts of dangerous comparisons by the Trump administration are over'.[15]

While in such statements the use of gas chambers was treated as a clear case of the use of chemical weapons, it can be noted that not everyone would subscribe to such classification. For instance, Spicer's contention that Hitler 'didn't even sink to using chemical weapons' would not at all be out of place in many academic and policy analyses of the history of chemical weapons.[16] As with the previously mentioned example of 'acid attacks', within such professional discourse, confined chemical chamber exposures are often treated as falling outside of the category of 'chemical weapons'. Instead this term is routinely taken to apply to situations of armed conflict. More generally, the twentieth century is replete with examples of disputes about what to include under the heading of chemical weapons – including many options used by the United States in wars, such as Agent Orange and napalm – rooted in differing criteria about what matters.[17]

A caution that follows from these observations is that the meaning of categories is dependent on who is speaking to whom and for what purposes. Rather than being treated as reflecting some pre-existing order, distinctions need to be treated as attempts at ordering. In this spirit, Etienne Aucouturier's chapter 'Biological Warfare, Chemical Warfare and the Public Body' sets out to revise the commonplace distinction in policy parlance that chemical and biological warfare (CBW) denote the uses of two different types of weapon systems: living organisms or products of living organisms versus inert, synthetic, chemical substances. His argument challenges this nomenclature by (re)connecting these weapon systems under a common definition. Aucouturier offers this argument to question the political projects behind modern warfare, specifically that target physiology.

But just as the definition of what counts as a chemical weapon needs to be thought of as provisional and potentially open to contestation, so too does their charge. It is not simply the case that the labels associated with the coercive use of chemicals carry negative connotations that foster moral aversion and thereby reticence to make use of them. For instance, while a term like 'tear gas' might frequently serve as a byword for severity in news headlines, this term arguably also masks the severity of effects. The name suggests an unpleasant experience from exposure, but also one limited in its effects. If this form of 'riot control agent' (RCA) were referred to as 'vomiting gas', a different set of connotations would be evoked that would likely call for an alternative set of justifications. So too if tear gas were labelled as a type of 'chemical weapon'. As noted in more than one chapter in this volume, arguments have been made for this designation since its development. More generally, conflict is characterized by language that serves to plaster over a sense of the human costs of what is at stake, as in terms such as 'collateral damage' or 'proportionality'. As a result, visibility and invisibility come together in complex ways through how distinctions are drawn and how labels are applied.

If wording needs to be approached with a sense of caution and nuance, so too with repugnance. While forms of moral repulsion are often associated with the hostile use of chemicals, arguably so too are forms of fixation and attention – in other words aversion can come accompanied by pull. The potential for such conflicting responses needs to be considered. As Miller has argued more widely than the topics of this book, 'The disgusting itself has the power to allure'.[18] While positing certain weapons in and of themselves as having some inherent power seems misplaced given the historical negotiations over meaning, it does seem justified to claim overall that chemical weapons today both disgust and fascinate simultaneously. In addition, it seems justified to claim that the more chemical agents are deemed exceptional, the greater extremes in the pulls and pushes.

To treat certain weapons as distinctive is not innocent in terms of its effects. As has been argued, the effort to stigmatize and eliminate chemical and biological weapons (CBW) across the globe has been highly advantageous for those major military powers seeking to simultaneously retain their nuclear weapons capacity.[19] Moreover, the way seeing and unseeing come together in categorical positioning of weapons today is highly complex. Groupings under which chemical weapons fall that are meant to separate them out as exceptional – such as 'WMD' and 'unconventional' weapons – can serve to legitimate or normalize the 'conventional' weapons that do not fall into the spotlight.[20] By keeping an eye on certain things rather than others, we are always producing a particular understanding of what is going on; the danger though is that this marginalizes important matters.

The aforementioned cautions mean that the trick in this volume is how to draw on both certain kinds of distinctions and categories while also questioning their meaning. The definitions of what counts as a 'chemical weapon', a 'chemical agent', a 'crowd control agent', 'coercion' and so on are treated as matters to be worked through for their commitments and implications, not objects of study that can simply be defined definitely from the start. Our primary interest lies in exploring the presumptions underlying debates about weapons intended to chemically affect humans – whether labelled as 'drugs', 'riot control agents' or otherwise.

Given this way of holding on to categories, in this volume, chemical agents are conceived as 'technologies of transgression'. The term 'transgression' is used to signal several interrelated characteristics of chemical agents:

- One of the reasons cited for why chemical agents are held with revulsion is how they affect the body. They can get on, permeate and grotesquely convert our outer skin. They can enter our orifices. They can cause what was hidden inside to come outside through vomiting, running sores and lesions. They can cause spasms by getting inside the body to affect our nervous systems. In short, they cross bodily barriers that maintain health.

- As developed already, the use of chemical agents is often positioned as exceeding the limits of acceptability. Political scholars have contended that social taboos associated with poisoning provide the historical origin for why chemical weapons have become regarded as abhorrent today.[21] Such moral condemnations have helped secure and been secured by international agreements such as the 'no-first-use' pact codified in the 1925 Protocol for the Prohibition of the Use in War of Asphyxiating, Poisonous or other Gases, and of Bacteriological Methods of Warfare (or 'GP'), an international treaty that prohibited 'asphyxiating, poisonous or other gases, and of all analogous liquids, materials or devices' and 'bacteriological methods of warfare'. Another major treaty is the wide-ranging prohibition on the development, production, acquisition, stockpiling, retention, transfer or use of chemical weapons established by the 1993 Chemical Weapons Convention (CWC). And yet, as developed already, the construction of such limits should not be taken as resolving once and for all what is appropriate by way of force. Instead they provide the terrain for which contests about what counts as a transgression are played out. For instance, when Italian forces used mustard gas against Ethiopia during 1935–1936, only muted international condemnation followed from the major powers at the time despite Ethiopia being signed up to the GP.
- Finally, transgression refers to the way the practice of developing and using chemical agents often confounds neat distinctions and boundaries. This sense of transgression recurs as a theme along multiple lines. As an example, chemical agents originally developed for one purpose have ended up serving others. In this vein, while 2-chlorobenzalmalononitrile (CS) gas (a form of tear gas) was initially introduced into the Vietnam War by the U.S. Army for limited riot control situations, it eventually became used in a wide range of settings.[22] As well, the knowledge and techniques necessary for producing countermeasures against chemical attacks can further the refinement of offensive weaponry. As still another dimension of transgression, in the manner that dosage is all that can separate a medicine from a poison, efforts to produce incapacitating agents as relatively humanitarian options can produce comparatively lethal agents.[23] In these and other ways, the lines drawn around chemical agents can become blurred and even vanish altogether.

ARTICULATIONS

The previous section assessed how distinctions are made. This one extends the previous analysis by elaborating how what is what depends on what is related to what.

Consider again the 11 April 2017 press briefing by Sean Spicer. As noted previously, in the media fallout from his initial remarks, a host of concerns were brought to bear by commentators as part of their proposals on how to make sense of the contention that Hitler 'didn't even sink to using chemical weapons'. Spicer himself would speak to what he judged to be at stake in his press briefing remarks by appearing on CNN and commenting:

> I was obviously trying to make a point about the heinous acts that Assad had made against his own people last week, using chemical weapons and gas. Frankly, I mistakenly made an inappropriate and insensitive reference to the Holocaust, for which there is no comparison and for that I apologize. It was a mistake to do that.[24]

In this statement, Spicer's initial attempt to place Assad even further beyond the pale than Hitler vis-à-vis the use of chemical weapons gets deemed unfortunate given the uniqueness of the Holocaust. At the time, media coverage cited other commentators (sometimes identified through labels such as 'Democrat' or 'Jewish') as casting the press briefing statement in far wider terms than just a simple mistake by a single individual. One way this was done was by linking Spicer's comments to alleged anti-Semitism in the Trump administration and even treating it as a form of Holocaust denial.[25] As suggested then, this example illustrates how the relevance and the meaning of categories formed by distinctions need to be appreciated as matters open for ongoing negotiation. As part of this, debate about an initial category can quickly spill out as further distinctions are drawn, thereby multiplying the demands of determining what is at hand. With this potential, questions of what context is necessary to settle the meaning of statements become highly germane. Is it appropriate to stay with the specific occasion specific words were uttered in order to understand what actors themselves treated as germane, or is it necessary to situate them further against a wider, external backdrop that can serve as explanation? The choices made can radically alter how situations and individuals are understood. Evoke a sense of a wider Trump administration anti-Semitism, for instance, and Sean Spicer moves from a person making a slip to one bound in a web of institutional and communal influences operating (more or less) behind the scenes. In turn, such a characterization would suggest not taking any of his statements about why he said what he said at face value. The question of where to look applies to many aspects of trying to piece together a story of Spicer's remarks. For instance, can one rely on selective quotes in news reports as definite building blocks or must these too be situated further (i.e. against other statements made by quoted speakers and information about their backgrounds)? If different types of information were to give grounds for different types of conclusions, where would one look to further help settle meaning?

In the specific case of the coverage of Spicer's press briefing remarks, meaning-making was particularly problematic because – while much controversy flared – the distinctions and evaluations being drawn were arguably not that well elaborated at times. It is possible for scholars or anyone else to venture a guess at what was implied by going beyond the specific terms and issues made explicit, but to do so begs the question about how this can be justified.[26]

In short, this example speaks to what Dilley referred to as the 'problem of context'.[27] He coined this phrase specifically to speak to the persistent and intractable dilemmas faced by anthropologists in trying to understand cultural phenomenon. For such scholars, and indeed for anyone else, a commonplace move for understanding deeds and words is to place them 'in context' – be that historical, social, cultural or any other set of considerations deemed pertinent. For Dilley, such a move raises profound and intractable questions such as those that follow:

- How can context provide the basis for determining meaning when what counts as 'the context' is often driven by interests, theories, questions and so forth? In other words, how can context be relied upon when notions of it result from prior forms of interpretation and, indeed, contextualization?
- How does contextualizing end? For instance, should the contexts marshalled to interpret some aspect of the world themselves be contextualized in order to understand their relevance? If so, when does this process stop or how could it be delimited in a manner to make to the notion useable?

Dilley traces such questions through long-standing debates in anthropology between relativism and realism, between cultural unity and uniqueness and between conceptualizations of language. Similarly, Asdal and Moser trace this notion through long-standing debates in science and technology studies (STS) regarding how to move beyond treating science and technology as products of some inner logic, while avoiding reducing them to being a mere 'a function of context'.[28] An upshot of such analyses is that context should not be regarded as some definitive thing 'out-there' that can be marshalled for explanation. Instead, reaching for it is a highly contingent move. Herein, context 'is both constitutive of social action and itself the outcome of social action; it is both a generative principle and the resulting outcome'.[29]

The question of what counts as the context is a highly relevant one for examining the coercive use of chemical agents. As in the case of Sean Spicer, the exceptional status of chemical weapons is often accomplished by juxtaposing them against something – in his case comparing Assad's use with Hitler's 'non-use'. Another common contextualizing move to justify exceptionality is to compare the effects of this category of weapons to others such as 'kinetic weapons', 'conventional weapons' and 'small arms'. Alternatively, in his historical examination of how chemical weapons became the site for all manner of financial and symbolic investment after World War I, Jenkins contrasts their widely promulgated horrors in the United States with the lack of regard to lynching murders taking place at the same time in the American South.[30] The novelty of Jenkins's contextualization within the arms control literature reminds us that any attempt to place the matters at hand in one context is simultaneously an undertaking that severs a connection to other possible ones.

For many social theorists, the problems of context cannot be definitively resolved, rather what is needed is a heightened sense of awareness about how contextualizing gets done. A term advanced for the examining the process whereby (dis-)connections are actively made is 'articulation', another being 'contexting'.[31] With its focus on the doing of contextualization, to examine articulation is a call to examine how context is fashioned, used and enacted. It is also to examine the considerations turning on matters of identity or purpose that bring forth as germane particular objects, events and circumstances.

As implied already, articulation not only applies to those who are studied, but those doing the studying. As will be illustrated in subsequent chapters, understanding the place of chemical agents today in coercion often entails making connections between the police and the military, between national boundaries and between global standards and local events. Thus, to attend to articulation also calls for self-reflection. Regard for this call is perhaps most prominent in the volume in the chapter by Rappert, Smith and Gould titled 'Opening Spaces through Exhibiting Absences: Representing Secretive Pasts'. It explores their reasoning behind two forms of representation of the former apartheid-era chemical and biological weapon programme in South Africa: a monograph titled *Dis-eases of Secrecy: Tracing History, Memory and Justice* and an exhibition named *Poisoned Pasts: Legacies of South Africa's Chemical and Biowarfare Programme*. While for both representations, the question is posed of what should be included within histories of Project Coast, Rappert, Smith and Gould also attend to what is not included in these histories due to the limits of what is known and what can be expressed. The challenge they take up is how to intertwine what is present and what is absent within histories in order to offer novel possibilities for representing secretive and contentious pasts.

As furthered by this chapter, the notions of contextualization and articulation are especially pertinent to the themes of this volume because they directly bear on how weapons based on chemical agents are conceived. For instance, as in the case of international inspectors seeking to find clandestine 'WMD' programmes, it matters greatly to what extent weapons are conceived as discrete objects. If they are thought about as 'shiny, pointy things',[32] then one set of determinations might be drawn from a particular set of evidence. If, in contrast, WMD programmes are deemed as consisting of a heterogeneous system of documents, chemical agents, precursors, know-how, equipment and physical infrastructure that amount to something more like a capability than a 'thing', then a different set of conclusions can be drawn from the same clues. As Vogel details in the case of U.S. intelligence on Iraqi weapon programmes in the buildup to the 2003 invasion, alternative contexts for understanding intelligence (technical, historical and political) can lead to markedly opposing arguments about what has been detected by way of 'WMD'.[33] The consequences of these arguments can be matters of life and death.

Too in the case of the South African apartheid-era chemical and biological weapon programme, from which documentary evidence indicating the production of bottles of cholera or anthrax-laced cigarettes was alternatively interpreted as a means for ending or saving life through the types of connections drawn.[34] As part of the positioning being fashioned, the interpretation of any object and context can be tightly bound together. What a bottle of cholera is, for instance, depends on the context against which it is situated, but the context gets defined too through the interpretation of a bottle. This suggests a dynamic and emerging state of affairs. Work in STS has employed the notion of 'co-production' to examine how these kinds of natural and social orders – knowledge and its social contexts – are produced together.[35]

Additionally, within disciplines such as STS, the term 'socio-technical systems' has been coined to signal the importance of conceiving of technology as a heterogeneous assemblage of many manner of things.[36] It takes a myriad of connections formed from supporting infrastructures, samples, equipment, software and so on for chemical weapons to function as advertised. To note this though is simultaneously to point out the potential for the unruliness of socio-technical systems. Each distinct element of a system might enable some goal to be achieved, but each too can act disruptively to the rest of the system and undermine the prospects of realizing the intended goal.[37] In the case of desired covertness, successfully keeping the chemical weapons programme hidden can require keeping the whole (or at least much of the) system hidden, a task that can prove formidable in practice.[38]

The concept of socio-technical system includes humans too. As will be developed in the subsequent chapters, treating bodies as part of more heterogeneous assemblages affords a range of new ways of seeing and understanding the effects, forms and configurations of weapons. This means acknowledging ways in which claims made about the 'non-lethality' of some types of chemical incapacitants ignore or sideline how individuals vary in their response to their usage. It also means acknowledging the vital role that can be played by expertise. In 'Tear Gas Epistemology: The Himsworth Committee and Weapons as Drugs', Balmer, Spelling and McLeish demonstrate how the characterization of CS gas was inter-twined with different forms of epistemic authority, each 'licenced' to make particular claims about these weapons. Experts such as scientists, lawyers, diplomats, civil servants and politicians, military commanders and the police all made claims, based on their own domains of expertise, about what tear gases 'are' and what therefore ought to be done with these weapons. Social studies of science have long recognized how science intersects with different forms of expertise, for instance, public, legal or military knowledge. Balmer and colleagues' study builds on this work and shows how tear gases had a large number of epistemic constituencies speaking on their behalf, thus providing a novel opportunity to understand other cases where numerous different expertises compete to make knowledge claims about weapons.

EXPERIMENTATION

Chemical agents have often been embraced as a means for sorting individuals and spaces, as well as resisted as unleashing forms of uncontrollable contamination. Both movements entail reference to experimentation. Building on the previous discussion of distinction and contextualization, this section asks how this notion provides an entry point for understanding how the use of chemical agents is to multiply a form of techno-politics.

Experimentation figures as relevant along various dimensions. One is the manner in which the development of weapons often involves forms of testing. Some recent scholarship on this issue has focused on cases of human experimentation, particularly in the context of controversies about scientific knowledge production as part of national security programmes.[39] In this volume, human experimentation figures in many chapters, as in the one by Guillemin extending her previous research on the notorious activities of the Japanese Army's Unit 731 during World War II.

One of the many issues in need of note in relation to experimentation is who counts as 'meriting' the status as a victim. History would suggest not all those affected are given the same status. Some categories of research subjects have been regarded as more noteworthy or more legitimate than others. For instance, as part of Project Coast, experiments of various sorts were conducted on enemy combatants, political opponents and dissidents.[40] Arguably, a group given far less attention in past investigations has been the member of the South African Defence Force (SADF). 'Volunteers' from Special Forces and 7 Medical Battalion Group were among those believed to have been subjects of human experimentation.[41]

Such contrasting levels of regard are not unprecedented. As has been contended in relation to the study of victimhood more widely, the 'ideal victim' is generally someone understood as weak compared to their offender, blameless for what transpired and a stranger to a clearly reproachable offender.[42]

In relation to chemical weapons, as in the case of President Trump's statement at the start of this chapter, being able to characterize those affected as innocent is often mobilized as part of underscoring their suffering and the need for action. So too is victimhood defined in relation to who is doing the harming. As contended by Christie, it typically follows that 'the more ideal a victim is, the more ideal becomes the offender. The more ideal the offender, the more ideal is the victim'.[43] In the case of the 2017 nerve agent attacks in Syria, the characterization of Bashar al-Assad as a dictator went hand in hand with the victim status of those affected. In contrast, in the case of Project Coast, those running the programme were respected members of their professions and held leading positions in science, a status which frustrated professional condemnation.[44]

A complicating consideration to the points of the previous paragraph is that any victim needs to be able to articulate his or her plight but also in a way that does not substantially threaten other interests.[45] During the CIA's 'mind control' programme titled Project MKUltra, high on the list of those covertly administered lysergic acid diethylamide in the 1950s included drug addicts, mental patients and prostitutes; groups that lacked an effective voice to contest.[46] In relation to the development of chemical weapons, one group that has repeatedly been a subject of experiments but whose troubles have been marginalized or ignored is that of (non-human) animals. While in theory, animals fit many of the characteristics associated with 'ideal victims', it has not followed that their suffering has been prominent in the history of chemical weapons. One reason for this has been the way those who have tried to give voice to animals have been labelled as 'unscientific, emotional and irrational activists'.[47] In other instances, those in official guardianship roles for animals have been circumvented. In the case of Project Coast, for instance, when animal experimentation was planned for a university site that the Society for the Prevention of Cruelty to Animals already had entry to through an unrelated lawsuit, a plan was hatched to designate the site as an 'animal hospital' instead of an 'experimentation unit'. This was devised as a plan because the former would not have afforded the same access rights. Today, what can be said about the suffering inflicted on animals from this programme – as with many others – derives from surviving research laboratory reports, documents that, by standard practice in science, employed an objectifying and passive language that removes traces of animal suffering.[48]

Experimentation does not stop at the lab door. Another dimension along which it proves a useful theme is how the deployment of chemical agents functions in practice. In other words, the street is also a field site. For chemical agents to have their desired effect – whether that be in terms of their lethality or non-lethality – depends on the level of exposures, who is exposed and what countermeasures are to hand. Especially in the case of agents claimed to be 'non-lethal', 'less-lethal', 'non-injurious' and so forth, a problem has been that the safety assumptions, formal rules, intellectual abstractions and defined boundaries presupposed have turned out to be ill-suited to messy and specific lived settings of use. In Aucouturier's chapter, his notion that chemical and biological warfare might be unified under an umbrella concept represents an attempt to address these ill-suited boundaries.

In his chapter titled 'What Counts as a Chemical Weapon?: The Category of Law Enforcement in the Chemical Weapons Convention', Michael Crowley explores the existing use and misuse of RCAs – tear gas, pepper spray and malodorants – by law enforcement officials through the world. Crowley details how the inadequate current regulation of such toxic chemicals has resulted in their widespread misuse, facilitating serious human rights violations including ill-treatment, torture, collective punishment and the suppression of freedom of expression, association and assembly. Further, his chapter explores how the current situation could worsen due to the ongoing development and marketing of new delivery systems capable of dispersing significantly greater amounts of RCAs over extended distances and large areas. Crowley then examines troubling research and development by certain states of weapons employing a distinct range of toxic chemicals, so-called incapacitating chemical agents (ICAs), purportedly intended to act on the body's core biochemical and physiological systems to cause prolonged but non-permanent disability, but which in inappropriate amounts can be lethal. Crowley examines the disputed legitimacy of such weapons, which have been promoted for (and in one controversial instance used in) large-scale anti-terrorist hostage operations but have been opposed by scientific and medical associations, arms control, human rights and humanitarian organizations and a growing number of states because of the grave threats they pose to health, human rights and human security.

Echoing points raised previously, some people and some spaces are more likely to be deemed legitimate subjects and sites for lived technological experimentation. Post World War I and intensifying after World War II, for instance, European colonies struggling for independence served as spaces in which RCAs (as they are now termed) could be trialed in ways beyond their use at home. Just as in the case of environmental harms, such locales often operated as 'spatial knowledge gaps', wherein ignorance about health effects was tolerated if not cultivated.[49] In 'Tear Gas and Colonial Bodies in the British Interwar Period', Anna Feigenbaum scrutinizes the cultural, economic and geopolitical aspects of British colonial administrators' decisions to take up tear gas as a means for maintaining control in their vast empire. The analysis begins in the 1920s, with the rise of tear gas as a means of controlling crowds in the post–World War I era. First adopted by the United States, France and South Africa, soon other imperial powers became curious about what tear gas might be able to do for them. As Feigenbaum argues, maintaining order in the crowded bazaars and heat of India, in the rural villages of Kenya or in the mining districts of North Rhodesia brought particular security challenges for colonial administrators seeking to maintain control over their populations, particularly when colonial authority was being contested. The rise of passive resistance, engagement of women in protest and local tensions around religion, ethnicity and tradition made policing colonial bodies 'between the wars' a rising issue for internal security. During this time, the British needed to keep the everyday social – and often profitable – operations of their colonial labourers functioning, while at the same time only using as much force as the public would tolerate. Bullets and blood brought unwanted attention from the media, flared local tensions and often caused objections back in England. Yet, unlike their U.S. counterparts, the British were still fond of following the GP that banned the use of poison gas after the devastating consequences of World War I. Exploring this crucial time in colonial administration, this chapter looks at the geopolitical decisions made by the British Empire as they came to officially permit the use of tear gas on colonial subjects abroad.

While in the context of the theme of this volume, the notion of experimentation conjures up many concerns regarding how individuals and groups serve as unwilling guinea pigs, and the lack of willingness to engage in experimentation can also be a source of criticism. For instance, in the case of pharmaceutical drugs, the accumulation of data about side effects does not stop with the end of clinical trials. Instead, the post-approval monitoring of effects is a vital component of ensuring that drugs are administered in proper dosages to the appropriate kinds of people. A lack of willingness on the part of those deploying chemical agents for coercive purposes to do the same has been one line of criticism. As detailed by Balmer, Spelling and McLeish, after the 1969 Derry riots in Northern Ireland, a British government committee was set up to investigate the health effects of the CS tear gas used. Although overall making a case for the general health acceptability of this tear gas for healthy persons in an open space, the Himsworth Committee noted a number of limits in scientific understanding. With this recognition, it argued that chemical agents should be regarded as being more akin to drugs than weapons. However, when police forces in England and Wales began trials of CS incapacitant sprays in 1996, the deficiencies in both the pre-approval testing and the post-approval monitoring served as sources of concern.[50]

In his chapter 'Controlling and Caring for Public Bodies: Civil Defence Gas Tests in World War II Britain', Alex Mankoo goes back further in history to show how coercion and care have come together – at least for certain populations – in experimentation. In this chapter, he documents how during World War II, Air Raid Precautions (ARP) departments around Britain carried out tear gas 'tests' – gas releases, sometimes surprise releases – in public spaces, such as town squares and shopping centres, with the objective of bringing 'home to the public the necessity for respirators to be carried'. These demonstrations were a part of a nationwide voluntary effort aimed at minimizing casualties among the public in the event of gas raids, and were held at the approval of (and by) local authorities. His chapter uses newspaper sources from the British Newspaper Archive and policy documents from the British National Archives to examine how tear gas was used to control and bring order to bodies and spaces in these civic settings, and how tests shaped notions of what constituted proper domestic behaviour in wartime. Mankoo shows how the use of tear gas in these tests was wrapped up in a socio-technical assemblage of care, through which the state aimed to secure and protect the national population body with gas tests on local populations. As such, he points out that these tests constituted a large-scale informal form of human experimentation that enabled civil defence authorities to simulate potential gas attacks while also encouraging certain public practices.

In pointing to how chemicals were wrapped up with a rhetoric of care, and as with other chapters in this volume, Mankoo attests to the importance of attending to how events are framed and re-framed as part of shaping how to make sense of what is at stake. This echoes themes from elsewhere. As Bonneuil, Joly and Marris argued in the case of genetically modified crops, 'Different definitions of the same problem designate different potential victims or concerned publics, construct different plots and precedents, and make links between localized and/or specific controversies to national or international issues that have wider appeal, thus transforming a local malaise into an injustice or scandal as defined by universal values and norms'.[51] So too in this volume.

In highlighting the scope for contest over how a given problem is designated – and indeed whether it is 'a' problem and what that problem is – the contributors detail many ways in which technological developments and political power are jointly constituted. We further do so by examining how these various situated networks of chemical control serve to marginalize, legitimate and reshape people, frames and forms of knowledges within particular conceptions of space. Through this, a number of questions come into view:

- Why do some cases of chemical weapons come to light, while others remain hidden? Who is involved in these processes, and who is represented by these decisions?
- Why does the use of some chemicals 'matter' more than the use of other chemicals (e.g. so-called non-lethal or less-lethal agents)? When do distinctions between these forms of weaponry get emphasized, evaluated or scrutinized?
- Whose bodies 'matter' more than others? How do certain bodies become more, or less, appropriate for chemical intervention?

- Who should be telling stories about the use of chemical weapons or chemical and biological weapons programmes?
- On whose behalf do chemical agents assert control? And on whom? And developing this, on whose behalf do narratives about chemical agents assert control, and on whom?

In doing so, *Chemical Bodies* explores the production of spaces that make the use of body-controlling chemicals (and technologies of control in general) possible and permissible. Human geography has recently extended similar kinds of analysis to drone attacks[52] and the militarization of policing and civil society.[53] This volume seeks to generate discussion in geography and related disciplines that centres on the growing role of chemical technologies in these geopolitical assemblages of control.

In the final chapter, '"Chemical Bodies" and the Future of Control', we as editors reflect further on how the notion of control permeates many aspects of the use of chemical agents. In particular, the chapter will address the range of notions of 'chemical bodies' presented across the book, and contemplate the value of thinking about these bodies as broader techno-political assemblages. In short, it asks: How do these notions of chemical bodies mutually inform one another? And, what implications does this broader approach have for the way in which scholars, policy makers and activists think about chemical agents more generally?

NOTES

1 Donald Trump, 'Statement by President Trump on Syria' (Mar-a-Lago, Florida: The White House Office of the Press Secretary, 6 April 2017).

2 See Barbie Latza Nadeau, 'Trump Just Discovered the Slaughter of Syria's "Beautiful Babies"?', *Daily Beast*, 17 April 2017, available at http://www.thedailybeast.com/articles/2017/04/07/trump-just-discovered-the-slaughter-of-syria-s-beautiful-babies; Robert Fisk, 'If Trump Cares so Much about Syrian Babies, Why Is He Not Condemning the Rebels Who Slaughtered Children?' *The Independent*, 17 April 2017.

3 Jane Onyanga-Omara, 'Why Acid Attacks Are on the Rise in Britain', *USA Today*, 27 April 2017.

4 Hayward Eleanor and Emily Kent Smith, '"It Felt Like My Face Was Eating Itself": Nightclub Acid Attack Victim, 21, Bares Her Scars and Relives the Horror as Police Reveal Ferne McCann's Wanted Boyfriend Arthur Collins May Be Hiding in Spain', *Daily Mail*, 21 April 2017.

5 Faye Kirkland, 'Acid Attack Hospital Admissions Have Almost Doubled in Last 10 Years', *The Guardian*, 30 September 2017.

6 Sarah Marsh, 'Surge in Acid Attacks in England Leads to Calls to Restrict Sales', *The Guardian*, 7 July 2017.

7 See Amnesty International, 'Bahrain: Fears of Further Violent Crackdown on Uprising Anniversary', 14 February 2017, available at https://www.amnesty.org/en/latest/news/2017/02/bahrain-fears-of-further-violent-crackdown-on-uprising-anniversary/.
8 Amnesty International, 'Bahrain: Human Rights on the Brink of Crisis' 18 January 2017, available at https://www.amnesty.org/en/latest/news/2017/01/bahrain-human-rights-on-the-brink-of-crisis/.
9 BBC News, 'Trump Jerusalem Move: Tear Gas at Lebanon US Embassy Protest', 25 May 2017, available at http://www.bbc.co.uk/news/world-middle-east-42298782; BBC News, 'Protest against Brazil's President Temer Turns Violent', 25 May 2017, available at http://www.bbc.co.uk/news/av/world-latin-america-40039111/protest-against-brazil-s-president-temer-turns-violent.
10 Jonathan Clifton, 'A Membership Categorization Analysis of the Waco Siege', *Sociological Research Online*, 14, no. 5 (2009), available at http://www.socresonline.org.uk/14/5/8.html.
11 Sean Spicer, 'Press Briefing by Press Secretary Sean Spicer the White House – Office of the Press Secretary', #36, 11 April 2017, available at https://www.whitehouse.gov/briefings-statements/press-briefing-press-secretary-sean-spicer-041117/.
12 Ibid.
13 David Smith, Ben Jacobs and Tom McCarthy, 'Sean Spicer Apologizes for "Even Hitler Didn't Use Chemical Weapons" Gaffe', *The Guardian*, 12 April 2017, available at https://www.theguardian.com/us-news/2017/apr/11/sean-spicer-hitler-chemical-weapons-holocaust-assad.
14 Ibid.
15 Ibid.
16 OPCW, 'Brief History of Chemical Weapons Use', available at https://www.opcw.org/about-chemical-weapons/history-of-cw-use/; J. P. Perry Robinson, 'Chemical Arms Control and the Assimilation of Chemical Weapons', *International Journal*, 36, no. 3 (Arms Control, Summer, 1981), 515–534; Jonathan B Tucker, 'The Future of Chemical Weapons', *The New Atlantis*, 26 (Fall 2009/Winter 2010), 3–29; Richard Price, 'A Genealogy of the Chemical Weapons Taboo', *International Organization*, 49, no. 1 (1995), 73–103.
17 David Zierler, *The Invention of Ecocide: Agent Orange, Vietnam, and the Scientists Who Changed the Way We Think about the Environment* (Athens: University of Georgia Press, 2011); Brian Rappert, *Controlling the Weapons of War: Politics, Persuasion and the Prohibition of Inhumanity* (London: Routledge, 2006), Chapter 5.
18 William Ian Miller, *The Anatomy of Disgust* (London: Harvard University Press, 1997), 111.
19 Richard Falk, 'The Challenges of Biological Weaponry', in *Biological Warfare and Disarmament*, ed. Susan Wright (London: Rowman & Littlefield, 2001), 29.
20 Brian Rappert, *How to Look Good in a War: Justifying and Challenging State Violence* (London: Pluto, 2012), Chapter 6.
21 Richard Price and Nina Tannenwald, 'Norms and Deterrence', in *The Culture of National Security*, ed. P. Katzenstein (New York: Columbia University Press, 1996), 114–152.

22 Malcolm Dando, *A New Form of Warfare* (London: Potomac, 1997), Chapter 5.
23 L. Klotz, M. Furmanski and M. Wheelis, 'Beware of the Siren's Song: Why "Non-Lethal" Incapacitating Chemical Weapons Are Lethal', March 2003, available at www.fas.org/bwc/papers/sirens_song.pdf; Michael Crowley, *Chemical Control: Regulation of Incapacitating Chemical Agent Weapons, Riot Control Agents and Their Means of Delivery* (Basingstoke: Palgrave Macmillan, 2015), 115–117.
24 Dan Merica and Elizabeth Landers, 'Spicer Apologizes for Hitler Comparison: "It Was a Mistake to Do That"', *CNN*, 12 April 2017.
25 Ibid.; Smith, Jacobs and McCarthy, 'Sean Spicer apologizes'.
26 For an analysis of the use force that engages with such questions, see Michael Mair, Chris Elsey, Paul V. Smith and Patrick G. Watson, 'Interpretive Asymmetry, Retrospective Inquiry and the Explication of Action in an Incident of Friendly Fire', *Symbolic Interaction*, 36, no. 4 (2015), 398–416.
27 Roy Dilly, 'Introduction', in *The Problem of Context* (Oxford: Berghahn, 1999).
28 Kristin Asdal and Ingunn Moser, 'Experiments in Context and Contexting' *Science, Technology, & Human Values*, 37, no. 4 (2012), 293.
29 Dilly, 'Introduction', 19.
30 Dominic Jenkins, *The Final Frontier: America, Science and Terror* (London: Verso, 2002).
31 Asdal and Moser, 'Experiments in Context and Contexting'.
32 J. Borger, 'Revelation Casts Doubt on Iraq Find', *The Guardian*, 7 October 2003.
33 Kathleen Vogel, *Phantom Menace or Looming Danger?* (London: Johns Hopkins University Press, 2012).
34 Brian Rappert and Chandré Gould. *The Dis-eases of Secrecy* (Johannesburg: Jacanda, 2018).
35 Sheila Jasanoff, *States of Knowledge: The Co-production of Science and Social Order* (London: Routledge, 2004).
36 Brian Rappert and Brian Balmer, 'Rethinking Secrecy and Disclosure: What Science and Technology Studies Can Offer Attempts to Governing WMD Threats', in *Technology and Security*, ed. B. Rappert (London: Palgrave, 2007).
37 Bruno Latour, *Science in Action* (Cambridge, MA: Harvard University Press, 1987).
38 Rappert and Gould, *The Dis-eases of Secrecy*.
39 Rob Evans, *Gassed: Behind the Scenes at Porton Down* (London: House of Stratus, 2002); Brian Balmer, *Secrecy and Science* (Abingdon: Ashgate, 2012); Ulf Schmidt, *Secret Science* (Oxford: Oxford University Press, 2015).
40 Rappert and Gould, *The Dis-eases of Secrecy*.
41 Chandré Gould and Peter Folb, *Project Coast: Apartheid's Chemical and Biological Warfare Programme*, UNIDIR/2002/12 (Geneva: UNIDIR, 2002), 84.
42 Nils Christie, 'The Ideal Victim', in *From Crime Policy to Victim Policy*, ed. E. Fattah (Basingstoke: Macmillan, 1986), 17–30.
43 Ibid.
44 Rappert and Gould, *The Dis-eases of Secrecy*.
45 Christie, 'The Ideal Victim'.
46 Tim Weiner, 'Sidney Gottlieb, 80, Dies; Took LSD to C.I.A.', *New York Times*, 10 March 1999, available at http://www.nytimes.com/1999/03/10/us/sidney-gottlieb-80-dies-took-lsd-to-cia.html.

47 For an example in relation to Project Coast, see Michele Pickover, *Animal Rights in South Africa* (Cape Town: Double Storey, 2005), 131.
48 Rappert and Gould, *The Dis-eases of Secrecy*.
49 S. Frickel and M. B. Vincent, 'Katrina's Contamination', in *Dynamics of Disaster*, eds. R. A. Dowty and B. L. Allen (London: Earthscan, 2001), 11–28.
50 Brian Rappert, 'Health and Safety in Policing', *Social Science and Medicine*, 56, no. 6 (2003), 1269–1278.
51 Christophe Bonneuil, Pierre-Benoit Joly and Claire Marris, 'Disentrenching Experiment: The Construction of GM – Crop Field Trials as a Social Problem', *Science, Technology, & Human Values*, 33, no. 2 (2008), 206.
52 D. Gregory, 'The Everywhere War', *The Geographical Journal*, 177, no. 3 (2011), 238–250; I.G.R. Shaw, 'Predator Empire: The Geopolitics of US Drone Warfare', *Geopolitics*, 18, no. 3 (2013), 536–559; Lucy Suchman, 'Situational Awareness', *Media Tropes*, 5, no. 1, 1–24.
53 R.M. Bernazzoli and M. Richelle, 'Power, Place and Militarism: Toward a Comparative Geographic Analysis of Militarization', *Geography Compass*, 3, no. 1 (2009), 393–411; D. Gregory, 'War and Peace', *Transactions of the Institute of British Geographers*, 35, no. 2 (2010), 154–186.

Chapter 2

From Reviled Poisons to State Arsenals: The Un(necessary) Proliferation of Chemical Weapons

Jeanne Guillemin

The introduction of chemical weapons (CW) in World War I was remarkable in both its reliance on basic science and its rapidity with which the use of this innovation spread. In a surprise mass attack, on 22 April 1915, at Ypres in Belgium, the German military released 167 tons of chlorine gas from a line-up of 5,700 pressure cylinders. Carried by the wind, the toxic emission passed in minutes to the Allied trenches and quickly killed or incapacitated thousands. This stunning assault provoked a vigorous arms race, like nothing in history for its engagement of industry, academia and the military.[1] Laboratory scientists in each belligerent nation – Germany, Britain, Italy, Russia, Austro-Hungary and later the United States – competed for increasingly potent chemicals, with the number of candidate agents reaching the tens of thousands.[2] Phosgene, a lung irritant like chlorine but six times more poisonous, became the Allies' preferred agent, while the Germans relied on the less-volatile diphosgene. The most important chemical agent proved to be mustard gas, a blistering agent (vesicant), while tear gas (lachrymators) and agents that caused sneezing (sternutators) were also added to arsenals and used in combination or alone.

Despite the development of successive generations of gas masks and protective blankets, enough soldiers were caught defenceless for poison gases to become a defining symbol of the war's horrors, adding blindness, burns and death by asphyxiation to the casualties caused by conventional weapons and disease.[3] In addition, after the war, doubt grew that chemical weapons, hazardous to transport and difficult to target, had significantly contributed to battlefield victories. In addition, uncontrollable environmental contamination had killed or injured civilian bystanders by the thousands, particularly in France. A legacy of moral revulsion and ambivalence about their military value soon troubled chemical weapons. Nonetheless, for decades, in nations large and small, chemical weapons programmes, intent on secret development, testing, production and stockpiling, were created and endured within larger military establishments, sustained by a few key advocates.

Informed by this wider backdrop of the twentieth century, this chapter examines how a secretive CW programme was created and fostered in Imperial Japan. In particular, it outlines the U.S. post-war protection of Japan from prosecution for its chemical weapons in China. This sheltering would have a lasting relevance well beyond the lives of those who contributed to the Japanese programme; it gave the American CW programme and others a legitimacy that fostered decades of proliferation of little discernible military value and ultimately of significant hazard to civilians and environments where hidden and outdated arsenals accumulated.

The argument in this chapter spans three distinct phases within the twentieth-century history of state CW programmes. After World War I, the League of Nations, created in 1920, heralded a new, optimistic era of disarmament that shaped important treaties and promoted norms against the state use of chemical weapons. In a decade, though, a second phase of growing nationalism began, undermining the idealism of disarmament and weakening restraints on chemical weapons development, production and use. This phase began in the 1930s and lasted through World War II, with Imperial Japan persistently violating established norms in its war with China. Notably, the major powers, preferring conventional arms, refrained from attacking each other with chemicals. Following World War II, chemical weapons programmes proliferated within the context of the Cold War, when both chemicals and biologicals were characterized as 'weapons of mass destruction (WMD)' on a threat level with nuclear arms.

PHASE ONE: DISARMAMENT GOALS AFTER WORLD WAR I

The cost of World War I in human lives, environmental destruction and political upheaval outstripped all previous conflicts.[4] When the war ended in 1918, international cooperation based on reason and law seemed the only way to prevent future catastrophes. In response, the 1920 creation of the League of Nations heralded a new, institutional approach to peace centred on the political resolution of state conflicts.[5] Article 8 of the League's Covenant affirmed that 'the maintenance of peace requires the reduction of national armaments to the lowest point consistent with national safety and the enforcement by common action of international obligations'.[6] Member states agreed to 'full and frank information as to the scale of their armaments, their military, naval and air programs and the condition of such of their industries as are adaptable to warlike purposes'.

Although the United States refused to join the League, its representatives maintained an active presence at its meetings and selectively supported international agreements that protected American interests or reflected popular opinion. One concern in the 1920s, for example, was Japan's rise as a naval power, with an advanced fleet that made it more militarily competitive and threatened Western trade interests in East Asia, especially in China.[7] Japan had already intruded into northeast China (Manchuria), in 1910 colonized Korea and after the war was given control of former German outposts in the Pacific.[8] In 1921, the U.S. State Department took the lead in persuading Japan, along with the United Kingdom, France, Italy and other states to limit warship tonnage and reduce the numbers of their destroyers and submarines.[9] From this effort came the agreement for an 'Open Door' policy that guaranteed China autonomy in negotiating trade and international relations in general.

In the Treaty of Washington of 6 February 1922, international disarmament was again a goal. A general prohibition of chemical weapons was included, which the United States approved. In language taken from the Treaty of Versailles, which had banned Germany's renewed use of chemical weapons, the treaty also underscored the legal precedents for the prohibition, which included the 1899 and 1909 Hague Conventions and their bans on the use of poison weapons in war, as they might evolve from industrial products.[10] Although the treaty never came into force, it paved the way for the 1925 Geneva Protocol (GP), which repeated the Washington text nearly verbatim, forbidding 'asphyxiating, poisonous or other gases, and of all analogous liquids, materials or devices' already 'justly condemned by the general opinion of the civilized world' and 'universally accepted as part of International Law, binding alike the conscience and the practice of nations'.[11]

Almost as an afterthought, the authors of the treaty included a parallel restraint on biological warfare (BW).[12] If Germany rearmed, its military scientists might conceivably invent anti-personnel germ weapons. Unknown to its Allies, in 1922, France began exploring germ aerosols and bombs to respond to a possible German attack on their shared border.[13] The Soviet Union was on the same track, exploring the offensive and defensive aspects of microbiology – a fact that influenced treaty delegates to extend the prohibition to 'the use of bacteriological methods of warfare'.[14]

The GP was widely embraced in the interwar years; forty-three nations became parties, with the British Empire, France, Italy and Germany, and the Soviet Union leading the way. It did not, however, offer a complete guarantee against proliferation. The option remained for state parties to develop, produce and use chemical and germ weapons if they believed (as did the French and the Soviets) that they might need to retaliate in kind to a first strike by an enemy power.

Although public antipathy for chemical weapons remained strong and the U.S. Army had little interest in supporting a chemical arsenal, the U.S. Senate refused to vote to consider ratification of the protocol. With vigorous lobbying, General Amos Fries, head of the Chemical Warfare Service (CWS), had persuaded key committee heads in Congress and outspoken chemical industry leaders that his programme, unbounded by international agreements, was necessary to counter a future German threat.[15]

In the interwar years, the potential of air power began driving the modern concept of total war, that is, a state's fight to full victory by moving beyond battlegrounds to annihilate the enemy's industrial base – its factories, rail and shipping lines and workforce. As early as 1921, the Italian General Giulio Douhet predicted that aeronautics would revolutionize warfare. 'Air power makes it possible', he wrote, 'not only to make high explosive bombing raids over any sector of the enemy's territory, but also to ravage his whole country by chemical and bacteriological warfare'.[16] For decades to come, the strategic vision of chemical and biological weapons (CBW) would be promoted by their advocates but remain technically beyond reach. Meanwhile, the targeting of enemy cities and manufacturing centres with conventional bombs and incendiaries was on the horizon, only awaiting the development of heavyweight aircraft.[17]

In 1928, in a grand but futile gesture towards peace, sixty-three nations, including Japan and the Soviet Union, signed the Kellogg-Briand Pact (also called the Treaty of Paris) which condemned the use of war to resolve international conflicts and pledged instead the recourse to 'pacific means'.[18]

PHASE TWO: THE PATH TO WORLD WAR II

By 1930, the power of the League as a peacekeeping operation began to decline. In September 1931, claiming that Chinese troops had attacked the Japanese Kwantung army at a railway junction near Mukden (Shenyang), the Japanese military mobilized to conquer the northeast area of China then referred to by non-Chinese as Manchuria.[19] China appealed immediately to the Council of the League of Nations, whose most influential members, especially the United Kingdom, had important commercial interests in both countries, but not, unfortunately, in Manchuria. From the beginning, the League members hesitated to become involved as long as China to the south was not disrupted.[20] Japan's explanation for its aggression, which violated the 'Open Door' policy for China, was that its troops were acting in self-defence against chronic violent attacks and that, given the disorder inherent in Chinese society, Japan's expanded presence would help Manchuria become a successful independent nation. With Japan's consent, five months after the Mukden incident, the League appointed the Lytton Commission to investigate the conflict in Manchuria on-site. Meanwhile, the Japanese continued with their military expansion, without fearing to aggress elsewhere in China. In January, in retribution for a Chinese trade embargo, they provoked an armed conflict in Shanghai that killed thousands of civilians and sent over 250,000 others rushing for safety to the city's International Settlement – an early portent of similar attacks on China to come.[21]

On 1 March 1932, after installing the Kwantung army as its ruling force, Japan declared the creation of the independent state of Manchukuo. Belatedly, on 1 October, the Lytton Commission released its report, which condemned Manchukuo as an illegitimate puppet state and advised a multilateral conference to demilitarize it.[22] Nothing was done until, at a February 1933 meeting in Geneva, the other League members unanimously voted for Japan's withdrawal from Manchuria. Japan's delegate to the League argued that this new acquisition was vital to his country's existence, and that the United States would be just as unwilling to give up control of the Panama Canal or the British to relinquish Egypt.[23] Resigning from the League, Japan then abrogated its naval pacts with the West and began a major build-up of its naval forces and army, while also tightening control of Korea and Taiwan and amplifying its repressive rule at home.[24] In this same time frame, Japan began shipping chemical munitions to Manchukuo, testing them on captive Chinese, and creating its base there for chemical weapons research and development, which also included human experimentation and bomb tests.[25]

World conflict prompted by ultra-nationalism was escalating in ways the League of Nations was unable to arbitrate or survive.[26] A transgressor was Japan, where militarists were winning over moderates in the struggle for Japan's destiny as the supreme power in East Asia.[27] In July 1937, the Japanese initiated war against China. It started with a clash between Japanese army units and Chinese garrison forces at the Marco Polo Bridge (in Chinese, Lukouchiao), located on the rail line between Beijing and the interior city of Hankow (Hankou). Although the incident was initially settled by an armistice, the Japanese army soon moved reinforcements from Korea and then added three divisions from the homeland. Emperor Hirohito and the cabinet approved, after the fact.[28] The Second Sino-Japanese War had begun, with the same expansionist goals as the first, in 1894–1895, when Japan won Korea from China's Qing dynasty. In 1937, after easily gaining control of Beijing, the Japanese battled for three months to win Shanghai. Then, moving inland, on 13 December, their troops captured China's capital, Nanjing, and embarked on a seven-week rampage of mass murders, sexual assaults, burning and looting that became known as the 'Rape of Nanking' or the Nanjing Massacre. League members and the United States protested, but none took action in China's defence.

Like the United States, Japan had rejected becoming a party to the GP, with the goal of developing of a modern, science-based arsenal, of which chemical weapons were an integral part.[29] In 1928, Japan built a major production facility on Okunoshima Island, near Hiroshima, mainly for mustard and tear gases. By then, its Army Institute of Scientific Research already included forty chemical weapons laboratories, twenty affiliated workshops and an annex staffed with hundreds of scientists and technicians.

In October 1937, China's delegates to the League protested Japan's use of chemical weapons (identified as mustard gas) on unprotected Chinese troops in armed conflicts between Shanghai and Nanjing.[30] The weakened League could only express sympathy and remind its members that 'the civilized world' had rejected chemical weapons. In 1938, the Chinese returned with additional documented accusations of Japan's chemical weapons. The GP, to which China had acceded in 1929, had been violated, but, as in the case of Italy's attacks on Ethiopia, League members did not rise to the mutual defence that was part of their commitment to other members.

In September 1939, with Nazi Germany's invasion of Poland, what would become World War II began, eventually pitting the Allied powers of the United Kingdom, the United States, France and the Soviet Union against the Axis powers of Germany, Italy and Japan. It brought even more death and destruction than World War I, leaving most major nations (except the victorious United States) bankrupt and their cities in cinders, with surviving populations around the world decimated, dislocated and starving.

PHASE THREE: THE COLD WAR AND THE LEGITIMATION OF WMD

In the spring of 1945, revolted by the war's atrocities and with victory in sight, the Allied commitment to a future of global peace led to the creation of the UN and the decision to hold international military tribunals (IMTs) to judge Axis leaders for their war crimes. But the fact that the Allies, especially the United States, had won the war using total war technologies, among them improved incendiary bombs, shifted attention to the bellicose intentions of the Axis powers and away from disarmament.

On June 25, in San Francisco, the final draft of the UN charter was unanimously approved and then signed by the members of the Security Council (the United States, Great Britain, France, the Soviet Union and China) and forty-five other nations, all resolving to protect succeeding generations from 'the scourge of war'.[31] In an echo of the League of Nations' disarmament ambitions, the UN goal of establishing 'a system for the regulation of armaments' was given a mandate in Article 26.[32]

In this cooperative atmosphere, the Allies also agreed that the leaders of the defeated Axis powers should be tried in international military courts. In

Washington, Secretary of War Henry Stimson planned for an Allied military tribunal to be conducted when the Third Reich was defeated.[33] Stimson's initiative, carried out by a small working group, produced the London charter, which mapped the goals and structure of an IMT to prosecute major Nazi war criminals.[34] With modifications, it also became the framework for the International Military Tribunal for the Far East (IMTFE) held in Tokyo following Japan's surrender.

In defining war crimes, the charter focused primarily on the wartime aggression of Axis leaders on peaceful nations. For this new category of 'crimes against peace', the charter's framers drew on the precedent of the Kellogg-Briand Pact and other pre-war treaties against the initiation of war. To address Nazi wartime murders of civilians, the authors of the charter introduced the concept of 'crimes against humanity', specifically, 'murder, extermination, enslavement, deportation and other inhumane acts committed against any civilian population'.[35]

The London charter also allowed the prosecution of conventional war crimes based on precedents established by treaties and laws. For example, the Allies were entitled to prosecute Axis abuses of prisoners of war expressly banned by the 1929 Geneva Convention and earlier accords. Established rules of war also forbade economic exploitation, the unwarranted destruction of infrastructure and the environment and failure to control the criminal conduct of troops. In theory, defendants might be held responsible for state violations of the 1925 GP or the bans on poison weapons in the Hague Conventions – except for the general Allied aversion to disarmament.

The Nuremberg charter was approved on 8 August 1945, just after the United States dropped its atomic bomb on Hiroshima and just before its bombing of Nagasaki. The bombings raised serious ethical questions but did not provoke universal condemnation. From the perspective of the victorious Allies, the Americans had forced Japan's surrender and, in addition to saving hundreds of thousands of Allied troops, had saved civilization from the Axis powers.[36] The historical impact of Truman's decision to use atomic bombs on Japan was profound. Putting a maximum value on total war strategy, the bombs upended the usual sequence of land warfare, destroying the enemy population centres in one blow, not as the result of protracted land or sea warfare.[37] This new capability meant that the United States as the world's lone nuclear power could threaten the sudden mass killing of enemy civilians without fear of equal retaliation.

In this immediate post-war atmosphere, to wage aggressive war – that is, to make the first strike against a peaceful state – was criminal, but new weapons technologies were prized by state militaries regardless of the moral implications that troubled League representatives in the 1920s. The nuclear age had begun, marked by the American assumption that strategic weapons were not only justified but vital for national security in the next war. In the spring of 1944, just before Germany surrendered, the United States organized the ALSOS mission to discover and appropriate nuclear or any other scientifically advanced weapons invented by the enemy and to secretly recruit scientists who could offer information, for example, about rocket science or the new nerve gasses (sarin, tabun and soman) invented by the Germans, who had made no important progress towards an atomic bomb.[38] Nor, despite wartime intelligence to the contrary, had they invested significantly in germ weapons. This move away from the moral judgement of weapons prompted a special rule to grant Nazi experts immunity from war crimes prosecution and, in hundreds of cases, employ them in Allied military projects.[39]

After Japan's surrender, it became possible to discover more about its modern arsenal. As a corollary to ALSOS, a team of U.S. experts prepared to conduct a Scientific Intelligence Survey accompanied General MacArthur to Japan to investigate any nuclear chemical and biological weapons its scientists might have developed.[40] Any discoveries of value were to be kept secret and forwarded for evaluation to the appropriate military programme officials. For chemical and biological weapons, this meant passing information to the chief officer of the CWS in Washington, which also oversaw the biological warfare programme.

In October 1945, the survey team reported its basic findings to Washington. The Japanese had made no progress in nuclear weapons and its chemical arsenal, although considerable, consisted only of conventional types of agents (e.g. mustard gas, phosgene, lewisite and tear gases) well known in the West. As for its germ weapons, it was suspected that the Japanese were hiding valuable information and that more could be learned from military physicians in Japan who had spent their careers in Manchukuo.[41]

In November, Major General Charles Willoughby, head of U.S. military intelligence in Tokyo (G-2), was tasked with the continuation of the survey's mission to exploit all information about Japan's wartime chemical and germ weapons. Willoughby had been MacArthur's chief of intelligence during the Pacific War, and at the war's end, he was placed within General Headquarters as one of two intelligence chiefs. The other was Brigadier General Elliott Thorpe, who was in charge of the Counter-Intelligence Section and Supreme Commander for the Allied Powers' (SCAP) Civil Intelligence Section, with much broader authority than Willoughby's over censorship, civic monitoring and the arrest of potential war criminals.[42]

Although devoted to MacArthur, Willoughby was less than enthusiastic about SCAP's mission to convert Japan into a democracy and saw merit in the rule of Imperial Japan; MacArthur jokingly referred to him as 'my lovable fascist'.[43] In his follow-up supervision of the Scientific Intelligence Survey, Willoughby consistently protected former military experts from public scrutiny. For example, to prevent public alarm, his office cooperated with the Japanese government in secretly destroying the Okunoshima Island factory, which had a grim history of accidental worker injuries and deaths, and dumping its munitions stockpiles. He commissioned the Tokyo office of CWS intelligence to put together a six-volume compendium of records of the Japanese chemical arsenal and attack plan documents, an overview that deliberately excluded potentially incriminating information.[44] Published on 15 May 1946, the summary volume of this compendium, on Japan's offensive and defence activities, fully accepted the claim by Japanese officials that no lethal chemicals had been used by their troops during the war in China, only tear, sneezing and vomiting agents.

Reinforcing this denial was the imprisoned former Prime Minister and Foreign Minister Tojo Hideki, who had also been chief of staff of the Kwantung Army. Tojo was interviewed on 2 April 1946 by Colonel John Beebe of CWS intelligence. Pressed as to what high authorities would have had to approve gas warfare against the Chinese, he replied, 'In such a decision, the cabinet would have had to have approved and the Emperor also'. But he asserted, 'Of course, no such decision was made'.[45] Tojo, who had no hesitation speaking with the press had sounded a warning. General MacArthur, convinced that Emperor Hirohito needed to be preserved as a figure of authority, was protecting him from war crimes prosecution, against the wishes of other Allies and high U.S. government officials. If made public at the Tokyo trial, Tojo's claim could be damning. Tojo further argued that the U.S. atomic bombings of Japan were a far worse violation of the rules of law than any chemical warfare – at a time when SCAP, fearing public reaction, was suppressing information on the two events. Tojo's denials of Japanese chemical warfare were supported by Beebe's interview with another prisoner slated to be a defendant, Field Marshall Hata Shunroku, leader of Japanese ground forces in China from 1937 onwards.

WAR CRIMES AND THE INTERNATIONAL MILITARY TRIBUNAL FOR THE FAR EAST

The Japanese Instrument of Surrender, signed on 2 September 1945, clearly urged the criminal prosecution of Japanese leaders, as a corollary to the IMT to prosecute Nazi leadership, which was about to begin in Nuremberg. The organization of the Tokyo trial was assigned by the Allies to General of the Army Douglas MacArthur, SCAP. In October, in the early weeks of establishing his General Headquarters, MacArthur created the Allied Legal Section (LS), which among other responsibilities collected the files of Japanese war crimes suspects. In late November 1945, in anticipation of the IMTFE, LS instituted the International Prosecution Section (IPS), over which the

American lawyer Joseph B. Keenan, appointed by President Truman as chief of counsel, would preside.

The IMTFE charter shared the basic framework laid out in the Nuremberg charter. From the beginning, the IMT set a high standard for the rapid dispatch of justice. In Nuremberg, representatives from the four participating nations (the United States, United Kingdom, France and the Soviet Union) hardly missed a day of work and were much aided by the 'Teutonic penchant for meticulous record keeping'.[46] Compared to Nuremberg, the IMTFE was a complicated global venture in which all nine signatories to the Japanese surrender agreement were entitled to participate: the Four Powers plus China, Australia, Canada, New Zealand and the Netherlands, with India and the Philippines later added.

Eager to participate, China's Associate Prosecutor Hsiang Che-chun (Xiang Zhejun) arrived in Tokyo in February 1946, weeks before other representatives. American officials, including President Truman and General MacArthur, accorded China much respect as an ally, giving its delegates priority in signing both the UN Declaration and the Japanese Surrender Instrument. Keenan, the IPS chief counsel, intended to start the prosecution with China's case, featuring Japan's 1931 and 1937 incursions as crimes against peace and the Nanjing Massacre as a crime against humanity.

China also planned to resurrect the claims of Japanese chemical warfare it first made to the League of Nations in 1937. While Japan was not a party to the 1925 GP, it had agreed to the 1899 and 1907 Hague Conventions and the Treaty of Versailles and their bans on the use of noxious poisons in war.

China's other grievance concerned a 1940–1941 series of Japanese plague attacks on four of its cities, an accusation the Allies dismissed at the time, but which the Chinese Division proposed as another crime against humanity.[47]

A high official in the Chinese judiciary but understaffed, Hsiang underestimated the burden of proof the prosecution had to bear.[48] The enduring upheaval from the war, complicated by the civil war with Communist Party forces, had made it difficult to assemble trial-worthy evidence. To assist China, IPS chief of counsel Keenan assigned a half dozen American lawyers from his staff to formulate its case. In mid-March 1946, in an extraordinary gesture supported by General MacArthur, Keenan sent two staff lawyers, Colonel Thomas Morrow and David Nelson Sutton, on a month-long trip to China to conduct witness interviews and collect original documents. Sutton became the principle organizer for the Nanjing Massacre charges and others involving Japan's economic exploitation of China, for example, its opium trafficking and environmental destruction. Despite a report from the Chinese Ministry of Health on the 1940–1941 plague attacks, Sutton concluded that evidence the Japanese had used a low-flying airplane to spray four targeted cities with infectious material was 'inconclusive' and lacked eyewitness testimony.[49]

Colonel Morrow, on the other hand, was successful in discovering Chinese officers who could testify first-hand about lethal chemical weapons attacks and even provide Japanese munitions collected from battlefields.[50] The U.N. War Crimes Commission, working with Chinese officials, had also recorded over a thousand incidents of chemical attacks.[51] Morrow debriefed Colonel John Beebe at CWS in Tokyo on the results of his month-long trip and gave his written report (dated May 5) to Keenan who passed it to SCAP, Willoughby and the commanding officer of CWS in Washington, Major General Alden Waitt.[52]

Waitt, a long-time CWS officer, was in a difficult position. With its arsenals left unused during the war, its strategic plans outclassed by nuclear weapons, and its post-war budget slashed, the future of CWS looked bleak. Continuing President Roosevelt's support of the GP, of which Waitt was a vocal critic, President Truman adhered to a 'retaliation only' interpretation of the use of chemical weapons. Following Army orders in 1945, in public, Waitt envisioned a purely defensive, research-oriented CWS; behind the scenes, he argued that chemicals would be essential in the next, inevitably total war – presumably with the Soviet Union. During the war, Japan's use of chemicals in China was an accepted fact – even Waitt believed it.[53] In 1942, in defence of China, Roosevelt had warned Japan to cease the attacks or face massive retaliation, a caution Japan took seriously. Morrow's case evidence for Japan's use of chemical weapons ought not to have been controversial.

But in 1946, the prospect of prosecuting Japan for chemical weapons at the Tokyo trial could destroy Waitt's ambition for a truly offensive U.S. programme. The graphic images of blistered victims, the Japanese use of hydrocyanic acid, the killing agent in Zyklon-B gas used to exterminate Jews, the environmentally toxic munitions left behind in China, the accidental deaths at the Okunoshima plant and other revelations would be ruinous for CWS post-war prospects.

On May 13, Morrow circulated a summary of charges, co-authored with another American lawyer, concerning overall Japanese military aggression in China.[54] The IMTFE indictment, submitted on 29 April to the court judges, already stipulated in its Appendix D: 'In the wars against China, poison gas was used'. In the 13 May draft, Morrow reinforced the case for Japan's illegal use of chemical weapons, naming a half dozen witnesses and document sources. Morrow had also found Japanese records showing that over 2,000 Chinese troops had been killed by chemical attacks, which added to reports from the U.N. War Crimes Commission of over a thousand incidents of chemical attacks.[55] The 13 May brief was forwarded to G-2 and to CWS.

In the previous months, after the retirement of General Thorpe in February, General Willoughby was given control of all SCAP intelligence offices, including those of commander-in-chief over war crimes suspects.[56] His authority to influence the IPS amplified, Willoughby extended it to General Geoffrey Marshall, the CWS chief intelligence officer in Tokyo, who, with the cooperation of Eugene Williams, took an active role in revising the 13

May charges of chemical warfare, effectively erasing them.[57] All the evidences that Morrow had assembled – the eyewitness testimony, hydrocyanic acid analyses, diaries, photographs, identified munitions and incident accounts – were eliminated, leaving only a vague reference to Japan's use of 'poison gas'. In their place, a revised 24 May brief cautioned that some facts about Japanese chemical warfare could be 'referred to and briefly used in evidence but not over-emphasized'. Inserting phrases taken from Beebe's interviews with Tojo and Hata, the new version discounted the recorded fatalities and backed away from any reference to the GP:

> It does appear that 1) gas was used only in emergencies and for the most part tear, sneezing, and vomiting gas was used and not the vesicants, 2) the amount of casualties inflicted on the Chinese as evidenced from their statistics was a very small proportion of the total casualties suffered by the Chinese during the war, which is well over 3,800,000 according to their own records, and 3) in their interrogations Generals Hata and Tojo refer to the fact that in the United States we have used poisonous gases such as sneezing, vomiting and tear gasses in labor disputes and General Tojo in his interrogation about gas, raises the question, 'How about the atomic bomb?' which he claims is a much more outrageous weapon of warfare than poisonous gas.[58]

As Tojo warned, prosecuting Japan's use of chemical weapons could ignite defence accusations about Hiroshima and Nagasaki as crimes against peace.

The 24 May case outline also downplayed China's initial complaint to the League of Nations about Japanese chemical weapons: 'The reference to the use of poison gas in this warfare appears to have been made in the form of a complaint by the Chinese to the effect that the Japanese army used gas in Shanghai three and four October 1937, but which the Japanese emphatically deny. We do not intend to offer this in evidence but merely invite the committee's attention'. General Marshall promised to provide Williams with the new CWS summary report on Japanese chemical weapons and excerpts from documents provided by a Chinese military office on gas warfare defence – information to support the argument that the Japanese army had used only tear gas and other non-lethal agents.

Pressure on the IPS also came from Washington. On 1 June, a Top Secret cable addressed to chief counsel Keenan arrived from the office of the War Department Chief of Staff, then General Dwight D. Eisenhower, with copies to General MacArthur, his chief of staff and G-2. It questioned whether IPS had considered U.S. Army rules of land warfare that protected its right to use chemical weapons, without strict adherence to the GP. 'Your views are requested', the message stated, 'whether prosecution of that sect[ion] will reverse US position'.[59] The implication was that, if successful, IPS prosecution of the

Japanese for the first use of chemical weapons as a war crime could cancel the U.S. Army's future latitude for the same option or even push it towards ratification of the GP, which not even President Truman saw as feasible as early signs of the Cold War were emerging.

Under U.S. influence, China's charges of chemical weapons against Japan disappeared from IPS case arguments, ostensibly in the name of national security but in reality to protect a narrow bureaucratic interest within the military to preserve the CWS. With its accusations of Japanese chemical weapons reduced to a 'complaint', the Chinese Division pursued prosecution for the more flagrant Japanese crimes resulting from 'aggressive war'.

General Willoughby followed this first intrusion of military intelligence into the court's IPS with others. In early June, having created a Central Interrogation office, he informed Keenan that, in conformity with War Department policies, 'no interrogations of Japanese or other enemy aliens will be performed in Japan by agencies of any country, except by prior authority of and under the direct supervision of the Assistant Chief of Staff [Willoughby], G-2 General Headquarters'.[60] IPS prosecutors could interview witnesses only with G-2 approval and in the presence of American intelligence officers and interpreters.

CODA

With the reputation of the CWS intact, James Patterson, the new Secretary of War, was free to become its advocate. In late 1946, CWS was upgraded to permanent status as the Chemical Corps, and its budget was soon increased to wartime levels. Its development of nerve gas and other weapons and the U.S. offensive germ weapons expansion under its aegis, kept secret from the public, justified the funding increase. Strategic scenarios of chemical and biological weapons – the annihilating raids envisioned by Douhet – were touted by both the U.S. and UK programme leaders. In 1948, with U.S. encouragement, the UN accepted the definition of nuclear, chemical and biological weapons as 'WMD'.[61] Both were linked to advanced sciences, as nuclear weapons were to physics, and were indiscriminate and therefore appeared potentially terrifying in their impact. After 1953, when Eisenhower became U.S. president, 'the retaliation only' restraint was dropped, giving the president the authority to determine first use. The UN definition, which ensured the survival of the Chemical Corps, also encouraged the proliferation of chemical and biological weapons programmes over the next forty years. As large advanced nations were emulated by lesser states, chemicals became known as 'the poor man's nuclear weapons'.[62]

NOTES

1 Olivier Lepick, *La Grande Guerre Chimique* (Paris: Presses Universitaires de France, 1998).
2 SIPRI, *The Problem of Chemical and Biological Warfare Volume I: The Rise of CB Weapons* (New York: Humanities Press, 1971), 26–58.
3 In his anti-war poem, 'Dulce et decorum est', Wilfred Owen, the British soldier killed in battle in 1918, made famous the image of a dying comrade drowning 'as in a green sea' of gas. The poem was published posthumously in 1920, as public reactions against chemical warfare were increasing.
4 Margaret MacMillan, *The War That Ended Peace: The Road to 1914* (New York: Random House, 2013).
5 David Kennedy, 'The Move to Institutions', *Cardozo Law Review*, 8, no. 5 (April 1987), 841–988.
6 The text of the Covenant is available through the Avalon Project at the Yale University Law School at http://avalon.law.yale.edu/20th_century/leagcov.asp.
7 Christopher Howe, *The Origins of Japanese Trade Supremacy: Development and Technology in Asia from 1540 to the Pacific War* (London: Hurst & Company, 1996), 268–297.
8 Andrew Gordon, *A Modern History of Japan: From Tokugawa Times to the Present* (New York: Oxford University Press, 2002), 177–178.
9 Three treaties resulted in the Five-Power-Treaty on naval disarmament; the Four-Power Treaty for the United States, France, the United Kingdom and Japan to clarify alliances; and the Nine-Power Treaty that added Belgium, the Netherlands and Portugal to support stability in the region for colonial powers. Japan and China signed the Shangtung (Shandong) Treaty, whereby Japan returned to China control of that province and its railway system (https://history.state.gov/milestones/1921-1936/naval-conference).
10 SIPRI, *The Problem of Chemical and Biological Warfare, Volume IV: CB Disarmament Negotiations, 1920–1970* (New York: Humanities Press, 1971), 46–47; Richard M. Price, *The Chemical Weapons Taboo* (Ithaca, NY: Cornell University Press, 1997), 78–98.
11 SIPRI, *The Problem of Chemical and Biological Warfare, Volume III: CBW and the Law of War* (New York: Humanities Press, 1973), 21–27.
12 Erhard Geissler, 'Biological Warfare Activities in Germany, 1923–1945', in *Biological and Toxin Weapons: Research, Development and Use from the Middle Ages to* 1945, eds. Erhard Geissler and John van Courtland Moon (New York: Oxford University Press, 1999), 91–126.
13 Etienne Aucouturier, *La guerre biologique: Aventures francaises* (Paris: Editions Matériologiques, 2017), 65–80.
14 Paul Walker, 'The 1925 Geneva Protocol', Harvard-Sussex Programme Occasional Paper, July 2016.
15 Frederic J. Brown, *Chemical Warfare: A Study in Restraints* (New Brunswick, NJ: Transaction Books, 2006), 45–96.

16 Giulio Douhet, *Command of the Air*, trans. Dino Ferrari (London: Faber & Faber, 1942), 6–7.
17 John Buckley, *Air Power in the Age of Total War* (Bloomington: Indiana University Press, 1999).
18 The International Treaty for the Renunciation of War as an Instrument of National Policy.
19 Edwin P. Hoyt, *Japan's War: The Great Pacific Conflict* (New York: McGraw-Hill, 1986), 81–102. Hoyt described the Japanese military as 'run amok'.
20 Arthur K. Kuhn, 'The Lytton Report on the Manchurian Crisis', *The American Journal of International Law*, 27, no. 1 (January 1933), 96–100.
21 Hoyt, *Japan's War*, 97–101; Marius B. Jansen, *The Making of Modern Japan* (Cambridge, MA: Harvard University Press, 2002), 584584.
22 'Lytton Report, 1932' United Nations Office at Geneva (UNOG) Registry, Records and Archives Unit, Printed Documents (League of Nations), 1932; Sir John Simon (Secretary of State for Foreign Affairs), 'The Lytton Report. Japan and the League of Nations' UK National Archives CAB 24/235, November 1932.
23 Stewart Brown (UPI), 'Japan Stuns World, Withdraws from League', *UPI Archives*, 24 February 1933, available at http://100years.upi.com/sta_1933-02-24.html. Six years later, Matsuoka would briefly serve as Japan's foreign minister, and in 1946, he would be indicted at the IMTFE for his role in promoting Axis cooperation and aggressive war.
24 Andrew Gordon, *A Modern History of Japan: From Tokugawa Times to the Present* (New York: Oxford University Press, 2002), 189–200.
25 Peter Williams and David Wallace, *Unit 731: The Japanese Army's Secret of Secrets* (New York: Free Press, 1989); Sheldon Harris, *Factories of Death: Japanese Biological Warfare, 1932–1945, and the American Cover-up* (New York: Routledge, 2002).
26 In 1933, Nazi-controlled Germany also quit the League, claiming that the League's World Disarmament Conference had acted with prejudice by denying it arms parity with France. Fascist Italy was the next major nation to abandon the League, after violating the Kellogg-Briand Pact. In 1934, under Prime Minister Benito Mussolini, Italy attacked Ethiopia with the intention of increasing its colonial holdings in Africa. The League attempted reconciliation between the two members, but to no avail. Starting in December 1935, in violation of the GP, Italy began using asphyxiating and poisonous gases on unprotected Ethiopian troops and civilians. Despite personal protests from Ethiopia's Emperor Haile Selassie and documented proof of gas casualties from Red Cross physicians, the League failed to implement effective military and trade sanctions against Italy or to act to deter further gas attacks. The protocol became a 'dead letter', a treaty in effect but ignored. The war ended in May 1936 with Ethiopia's defeat and its incorporation into Italian East Africa, after which Italy exited the League. Meanwhile, Germany and Japan signed the Anti-Comintern Pact to solidify their alliance against international communism and the Soviet Union; Italy soon joined the agreement.
27 Jansen, *The Making of Modern Japan*, 576–585.

28 Herbert P. Bix, *Hirohito and the Making of Modern Japan* (New York: Harper-Collins, 2001), 318–327.

29 Walter E. Grunden, *Secret Weapons and World War II: Japan in the Shadow of Big Science* (Lawrence: University Press of Kansas, 2005), 178–182.

30 SIPRI 1971, *Vol. IV*, 189–192.

31 '1945: The San Francisco Conference', *United Nations*, http://www.un.org/en/sections/history-united-nations-charter/1945-san-francisco-conference/index.html.

32 Frits Kalshoven, 'Arms, Armaments, and International Law', *Collected Courses of the Hague Academy of International Law*, 191 (1985), 309–311.

33 Luc Reydams and Jan Wouters, 'The Politics of Establishing International Criminal Tribunals', in *International Prosecution*, eds. Luc Reydams, Jan Wouters and Cedric Ryngaert (New York: Oxford University Press, 2012), 6–80, 15; Mark J. Osiel, 'In Defense of Liberal Show Trials – Nuremberg and Beyond', in *Perspectives on the Nuremberg Trial*, ed. Guénaë Mettraux (New York: Oxford University Press, 2008), 704–726.

34 Telford Taylor, *The Anatomy of the Nuremberg Trials* (New York: Alfred A. Knopf, 1992), 34–42.

35 Nuremberg Charter, Section 6.

36 Thomas B. Allen and Norman Polmar, *Codename Downfall: The Secret Plan to Invade Japan – and Why Truman Dropped the Bomb* (New York: Simon & Schuster, 1995), 224; Douglas J. McEachin, 'The Final Months of the War with Japan: Signals Intelligence, U.S. Invasion Planning, and the A-Bomb Decision', CIA Center for the Study of Intelligence (CSI) Report no. 98–10001, 1998, available at https://www.cia.gov/library/center-for-the-study-of-intelligence/csi-publications/books-and-monographs/the-final-months-of-the-war-with-japan-signals-intelligence-u-s-invasion-planning-and-the-a-bomb-decision/csi9810001.html.

37 Thomas C. Schelling, *Arms and Influence* (New Haven, CT: Yale University Press, 2008), 19. 'Nuclear weapons', Schelling observed, 'made it possible to do monstrous violence to a people without first achieving victory'.

38 Boris Pash, *The Alsos Mission* (New York: Charter Books, 1969); Samuel A Goudsmit, *Alsos* (New York: Henry Schuman, 1947).

39 Clarence G. Lasby, *Operation Paperclip: German Scientist and the Cold War* (New York: Athenaeum, 1971); Linda Hunt, *Secret Agenda: The United States Government, Nazi Scientists, and Project Paperclip, 1944–1990* (New York: St. Martin's Press, 1991); Annie Jacobsen, *Operation Paperclip: The Secret Intelligence Program That Brought Nazi Scientists to America* (New York: Little, Brown, 2014).

40 Karl Taylor Compton, 'Mission to Tokyo', draft for *Report for Scientific Intelligence Survey in Japan 1945*, MIT, Office of the President, Personal Papers of Karl T. Compton and James R. Killian n.d., box 56, folder 15.

41 'Mission to Tokyo by Karl T. Compton', Office of the President, 1930–1959 (Karl Compton), MIT Archives, box 59, folder 15, 2–3, 8 October 1945.

42 Takemae Eiji, *Inside GHQ: The Allied Occupation of Japan and Its Legacy*, trans. Robert Ricketts and Sebastian Swann (New York: Continuum, 2002), 163.

43 Jeanne Guillemin, *Hidden Atrocities: Japanese Germ Warfare and American Obstruction of Justice at the Tokyo Trial* (New York: Columbia University Press, 2017), 67–68.
44 *Intelligence Report on Japanese Chemical Warfare Vol. I* (Tokyo: Office of the Chief Chemical Officer, GHQ, 15 May 1946), available at the Combined Arms Research Digital Library, http://cgsc.contentdm.oclc.org/.
45 Interrogation of General Hideki Tojo, NARA, RG 496, entry 53, box 342, 4, 2 April 1946.
46 Taylor, *The Anatomy of the Nuremberg Trials*, 56.
47 P. Z. King, 'Bacterial Warfare', *Chinese Medical Record*, 61, no. 3 (July–September 1943), 259–263.
48 Barak Kushner, *Men to Devils, Devils to Men: Japanese War Crimes and Chinese Justice* (Cambridge, MA: Harvard University Press, 2016), 77–84.
49 David Nelson Sutton to General Douglas MacArthur and Joseph B. Keenan, 'Bacteria Warfare', University of Richmond Law Library, Sutton Collection, Personal and Working Series, old folders 1–60, folder 13, 25 April 1946. See also Guillemin, *Hidden Atrocities*, for how G-2 also prevented prosecution of defendants for germ warfare against China.
50 Morrow as a combat soldier had witnessed the horrific German use of multiple chemicals at the 1918 Battle of Meuse.
51 '[United Nations] List of War Crimes', IPS Doc. 1700, University of Virginia Law Library, Tokyo War Crimes Trial Collection, sec. 7, folder 23, MSS1 Su 863 a, n.d.
52 Colonel Thomas H. Morrow, 'Japanese Use of Chemical and Biological Weapons against China, 5 May 1946, NARA, RG 496, entry 53, 343. Noted as from Keenan to GHQ, G-2, and chief chemical officer.
53 Alden H. Waitt, 'Poison Gas in This War', *New Republic*, 106, no. 17 (April 27, 1942), 563–565.
54 Mr. Kenneth N. Parkinson and Colonel Thomas H. Morrow to Mr. Joseph B. Keenan, chief of counsel, 'Trial Brief-All China Military Aggression', University of Virginia Law Library, Tokyo War Crimes Trial Collection, box 2, folder 12, MSS 93–4, 13 May 1946.
55 '[United Nations] List of War Crimes', IPS Doc. 1700, University of Virginia Law Library, Tokyo Trial Collection, sec. 7, folder 23, MSS1 Su 863 a, n.d.
56 Takemae, *Inside GHQ*, 164–168.
57 Mr. Kenneth N. Parkinson and Colonel Thomas H. Morrow, 'All China Military Aggression, 1937–1945', University of Virginia Law Library, Tokyo Trial Collection, box 3, folder 4, MSS 93–4, 32, 24 May 1946.
58 Ibid.
59 WARCOS to CINCAFPAC (FOR KEENAN IPS), ACTION: INTL PROSEC SECT, MacArthur Archives, RG-9, box 101, folder WD, 1–10 June 1946, 1 June 1946.
60 Chief of Counsel to Staff, 'Interrogations', University of Virginia Law School, Tokyo War Crimes Trial Collection, box 3, folder 1, MSS 78–3, 4 June 1946.

61 W. Seth Carus, 'Defining "Weapons of Mass Destruction"', Occasional Paper No. 8, Center for the Study of Weapons of Mass Destruction, National Defense University (Washington, DC: National Defense University Press, 2012), 9–11.

62 Jeanne Guillemin, *Biological Weapons: From the Invention of State-sponsored Programs to Contemporary Bioterrorism* (New York: Columbia University Press, 2005).

Chapter 3

Lesser Appreciations: A History of Interwar Chemical Warfare

James Revill and Marcos Favero

This chapter examines the role of chemical weapons (CW) in punishing, terrifying and demoralizing insurgents and 'colonials'. Drawing on the lesser known history of chemical weapons use during the interwar years by states against insurgent groups in the Tambov, Morocco, Ethiopia and China, the chapter shows that the applications of lethal chemical technology has been applied to eliminate discordant voices, frequently facilitated, it would seem, by deep convictions of racial or class superiority.

We begin by outlining the emergence of early prohibitions against the use of poison, focusing on how such prohibitions excluded those outside of 'an emerging society of civilized states'.[1] The chapter proceeds to explore theories of identity and othering, drawing on the work of Tajfel, Brewer and Bandura among others, to look at how ingroup/outgroup dichotomies can facilitate extreme acts of violence against an enemy, as well as enable individuals to eschew personal moral responsibility for the perpetration of such acts.

The chapter argues that the process of othering may create conditions more conducive to the use of chemical weapons, something explored through a number of micro-case studies of chemical weapons use by powerful states against what are loosely termed 'insurgent' actors. These micro-case studies are not intended to form a complete history of the identified instances of chemical weapons use by states – considerably more has been written on each of these episodes by scholars and researchers around the world.[2] Nor is this section intended to comprehensively cover all events over the course of this period; others could have been included, not least the alleged – and somewhat discredited claim – of British use of chemical weapons in Mesopotamia.

Nonetheless, looking at incidents that are perhaps less well known among Western audiences illustrates a degree of continuity in chemical weapons use over the interwar years and allows themes to be teased out from these asymmetrical chemical campaigns. The concluding section seeks to distil some of these themes. It draws attention to the significance of events and strategic shocks in justifying the use of chemical weapons against weaker and unprepared victims, and, more significantly, it considers the possible role of 'othering' as a facilitator of chemical weapons usage by perceived ingroups against outgroups. It is argued that such an analysis of the past can perhaps shed some light on present uses of chemical weapons, as well as potentially inform consideration of how and when chemical weapons may be used in the future.

OF 'CIVILIZED' POWERS AND 'OTHERING'

It has often been remarked that poison and later chemical weapons have been subject to a long-standing and cross-cultural taboo, and, indeed, there are a number of early condemnations of the use of poison[3] which have evolved into more formalized prohibitions with the emergence of the Westphalian international system. As Price notes, the chemical weapons taboo has been 'reinforced in the international realm by the appeal to civilization and civilized conduct'.[4]

As much is evident in relation to poison and gas warfare in the 1868 Declaration of St. Petersburg, the Hague Declaration of 1899 and the 1925 Geneva Protocol (GP). The 1868 Declaration of St. Petersburg was the product of the assembly of an International Military Commission, 'to examine the expediency of forbidding the use of certain projectiles in time of war between civilized nations' under which contracting parties renounced in 'war among themselves . . . any projectile of a weight below 400 grammes, which is either explosive or charged with fulminating or inflammable substances'.[5]

Similarly, the 1899 Hague Declaration on the Use of Projectiles the Object of Which Is the Diffusion of Asphyxiating or Deleterious Gases was binding only upon the Contracting Powers; with those contracting powers, as Price notes, constituting 'members of an emerging society of civilized states'.[6] Similar sentiments are also evident in the 1925 GP, which states the 'use in war of asphyxiating, poisonous or other gases, and of all analogous liquids, materials or devices, has been justly condemned by the general opinion of the civilized world'. The Protocol correspondingly binds high contracting parties; however, many states submitted reservations indicating the protocol was binding only upon such (civilized) parties and did not apply to those outside the civilized world.

As the introductory chapter to this volume notes, such distinctions matter as they 'cordon off parts of the world from each other'.[7] This is not something unique to chemical weapons: it is clearly manifest in the often cited prohibition emerging from the Second Lateran Council in 1139 on the use of the Crossbow against Christians, a prohibition that did not extend to non-Christians.[8] However, chemical weapons perhaps particularly neatly illustrate the 'demarcation of distinct zones of warfare' and how 'war among the 'civilized' advanced industrialized countries was a different matter from conflicts involving other areas such as the Colonies'.[9]

This distinction at the international level between appropriate conduct in wars against civilized actors, on the one hand, and what was acceptable in wars against others, on the other hand, has parallels at the national level. In the case of the British, Chapter XIV of the 1914 *Manual of Military Law* states:

> It must be emphasized that the rules of International Law Warfare apply only to warfare between civilized nations, where both parties understand them and are prepared to carry them out. They do not apply in wars with uncivilized States and tribes, where their place is taken by the discretion of the commander and such rules of justice and humanity as recommend themselves in the particular circumstances of the case.[10]

Such logic was seemingly manifest in the consideration of using chemical weapons by the British against tribes in Mesopotamia in the 1920s: as Churchill infamously stated, 'I do not understand this squeamishness about the use of gas. I am strongly in favour of using poisoned gas against uncivilized tribes'.[11] Several scholars have suggested that the British forces subsequently employed chemical weapons against Arabs in the region with a view to suppressing the rebellion.[12] However, others have countered this claim indicating that 'no such weapons existed'[13] and, accordingly, claims of mustard gas use appear 'to confuse what Churchill wished with what was practical at the time'.[14]

While the British case therefore appears to be a non-event, it is nonetheless illustrative of the application of the civilized/uncivilized dichotomy in considerations around using chemical weapons, and, as the subsequent sections indicate, several others states clearly did use chemical weapons against ill-prepared 'uncivilized' insurgent movements in the interwar years.

Such use appears to have been facilitated in part by othering, a process 'through which identities are set up in an unequal relationship'[15] and based on the establishment of what Sumner first termed 'ingroups' and 'outgroups'.[16] According to widely accepted social-psychological theories, identity is formed by a system of beliefs that are developed by individuals to reduce uncertainty,[17] foster belongingness,[18] cope with the overload of information[19] and make sense of a seemingly chaotic and volatile world.[20] Through a process of abstraction, personal identity constructs are formed and informed through perceptions of the world,[21] including social categorizations of identity based on perceived difference between one's ingroup and the outgroup of the other. As part of this process, there is a tendency to minimize the perceived differences in the ingroup and emphasize the differences between outgroups, leading to ingroup favouritism and a perceived sense of superiority over an outgroup.[22]

In a context of intergroup conflict, and in addition to making sense of the world, the manifestation of ingroups and outgroups can serve to justify aggression[23] through a process of moral disengagement that frees the individual from the cognitive dissonance generated by acknowledging certain negative actions against outgroups, allowing a disengagement of self-sanctions.[24] This moral disengagement can be achieved through a self-defence mechanism that eases the feeling of injustice by dehumanizing the outgroup as a social category that does not accept norms or values and, therefore, does not deserve human treatment.[25]

This homogenizes and deindividuates the outgroup as an evil entity deprived of humanity and morality, a faceless enemy that inspires negative emotions and provides a psychological permit to harm the other.[26] The dehumanization of the outgroup happens through epithets such as 'uncivilized' or 'primitives', or other zoomorphistic or somehow demonizing terms. In some cases, this can be furthered by characterizing the outgroup as violators of social norms, and therefore undeserving of them. Not only can the process aid in the justification of extreme acts of violence, the process of stripping the outgroup of its humanity also helps in minimizing any self-censure or guilt that might otherwise arise from acts of violence.[27] The following cases provide some insight into how ingroups and outgroup distinctions were manifest in chemical weapons use.

THE BOLSHEVIKS IN THE TAMBOV GOVERNORATE 1921

The first micro-case study concerns the reported Bolshevik use of chemical weapons against the peasants in the Tambov Oblast in 1921 as part of an attempt to suppress an uprising which, at one point, appeared to threaten the nascent Bolshevik regime. The peasants of the Tambov agricultural region, which lies some 400 kilometres to the southwest of Moscow, suffered during the Russian Civil War with *Prodrazvyorstka* or enforced food requisitioning practised by both sides. The increase in food requisitioning demands imposed upon the Tambov in 1920,[28] combined with the brutality of Bolshevik repression, led to a series of violent clashes between peasants and Bolshevik authorities that culminated in the peasants forming an organized resistance movement led by Alexander Antonov, a former police officer, under the banner of the Union of Toiling Peasants.

According to Figes, 'The speed at which the revolt spread caught the Bolsheviks in the Tambov unprepared'[29] and, more significantly, shocked Lenin, who reportedly acknowledged that the revolt was the 'greatest threat his regime had ever had to face'.[30] In the face of both a hostile military force numbering in the thousands and 'a hostile population of millions', Bolshevik rulers sought to quell the uprising as quickly as possible. However, initial efforts at repression in late 1920 and early 1921 were clumsy and unsuccessful, and it soon became apparent that conventional military force was poorly suited to dealing with the unconventional methods of the peasant movement.[31]

The oppression of peasant movements sits rather awkwardly with Marxist-Leninist ideological thinking. In order to accommodate such brutality, the conflict became framed as a class war against the *Kulaks*, that is, the more prosperous land-owning farmers, 'whose fate it was to be liquified as a class'.[32] In reality, the majority of Tambov peasants involved in the rebellion hardly fit the category of *Kulaks*; as Statiev has stated, 'Most of the peasants who resisted grain confiscations were in fact *seredniaks*',[33] or 'peasants of average prosperity'.[34] Nonetheless, the Soviet leadership expediently declared the insurgent regions *Kulak*, thereby engineering an outgroup, and 'persuaded themselves that the elimination of *Kulaks* would automatically end the insurgency'.[35]

This process of othering arguably provided both the ideological justification for more oppressive measures and a means of circumventing moral constraints. In April 1921, Lenin appointed General Tukhachevsky to lead the repression of the Tambov Province.[36] With nearly 100,000 troops and extensively equipped, Tukhachevsky waged a campaign of violent repression characterized by cruelty and terror[37] to deal with what he reportedly viewed as an ' "Epidemic" that infected the villages of the province' and which required a 'shock campaign . . . to annihilate the disease'.[38] The nature of the shock campaign is exemplified in Order No. 171 of 11 June 1921 which was signed by Tukhachevsky and instructed the troops to, inter alia, execute the eldest son of families found to be harboring arms or bandits.[39]

This practice of exemplary terror appears to have been a military success to some extent; however, it failed to deal with the remnants of the rebel forces located in the forest of the Tambov who were proving difficult and dangerous to attack.[40] Indeed, Bobkov suggests the forests in the North of the Tambov, from which the rebels were able to launch incursions, had long proven difficult to attack,[41] and it was the risks posed to Bolshevik forces by such operations that led to the search for less conventional means of quashing the rebellion.[42]

It is under these circumstances that by early June 1921, the decision was made to turn to gas.[43] While this force multiplication function was clearly important, it is also likely that the psychological value of such weapons played a role.[44] Specifically, the use of chemical weapons could be seen as demoralizing the Antonov movement as well as serving to terrorize local sympathizers. Accordingly, on the 12th of June, an order of the Tambov Province command was issued that authorized the use of poison gas in order to 'purge the forests where the bandits are hiding'. This was accompanied by a request to Moscow for chemical expertise and munitions.[45]

This request for gas weapons was not met with overwhelming support in Moscow, in part because of concerns over the capricious nature of gas warfare.[46] Nonetheless, a number of chemical munitions – the exact number and nature of which has been disputed – were duly dispatched to the Tambov along with instructions and specialist personnel.[47] Whether they were used, and, if so, the extent of use remains contested. As Landis notes, few primary records exist and none of them quite correspond generation of 'gas clouds' to clear the forests, as envisaged by Tukhachevsky, yet alone the more outlandish claims that the Tambov Command established death camps where prisoners were gassed.[48] What is acknowledged in one of the more detailed accounts of the event by Alexander Bobkov of the Russian Academy of Sciences is that there were at least two events in which chemical shells were used: first in Smol'naia Vershina on the 12th of July and second on the 2nd of August in Sukhie Dubki. The primary results of these attacks remain unclear but appear to have been limited to affecting the morale of the Antonov movement.[49]

THE SPANISH DURING THE RIF WAR 1921–1927

The second micro-case study concerns the use of chemical weapons by Spanish colonial forces against the Rif tribespeople under Abd el-Krim in Northern Morocco. For many years, the Spanish had cooperated with the tribes of the Rif; however, as Flemming notes, this cooperative relationship ended upon realization in 1919 that the Spanish intended to occupy swathes of Northern Morocco. Tension increased over the ensuing couple of years with a small number of guerrilla attacks on Spanish forts, culminating in the Annual disaster of 1921 in which some 8,000–12,000 Spanish soldiers were killed.[50]

Spanish interest in chemical weapons predated the Annual disaster, with some evidence indicating that requests had been made to Germany for 'samples and information about [CW] manufacture'[51] as early as 1918. However, the Annual disaster that appears to have 'awakened' more widespread Spanish interest in chemical weapons, converting 'what might have been considered a weapon of last resort into what they hoped would be the principal weapon for the repression of colonial rebels'.[52] As much was indicated by the Spanish high commissioner of Spanish Morocco (as well as some Spanish ministers)[53] in 1921 who indicated that despite previous reservations, after the 'treacherous and deceptive conduct' of the Rif tribes (in the Annual attack), he would use the gas with 'true fruition'.[54]

As was the case with the Tambov mentioned earlier – and the cases given further – the use of chemical weapons took place in the context of potent othering that served to justify aggression, something heightened following the Annual massacre. Indeed, King Alfonso XIII himself was reported as saying the important thing was 'the extermination, like that of a malicious beast, of the Beni Urriaguels and the tribes who are closest to Abdel Krim',[55] and Balfour suggests that over the course of the 1920s, there emerged an 'extreme right wing discourse increasingly shared by Alfonso's coterie of militarists'.[56]

This process of othering was reinforced by the frustration of the Spanish inability to succeed with conventional forces (and the success of the Rif tribesmen) and appears to have 'broke[n] down the moral barriers' against the use of chemical weapons,[57] something perhaps further facilitated by the humanitarian framing of chemical weapons as less harmful when compared to military action by shock troops.[58] Moreover, moral codes seemingly remained but applied only to the ingroup and excluded those perceived uncivilized groups with whom they were fighting.[59] In doing so, this process of othering stripped away the moral responsibility of the individual's violent acts, something facilitated by domestic censorship which mitigated any possible popular rebuking of Spanish forces by Spanish citizens. As one of the first pilots to drop mustard gas in Abyssinia later remarked:

> We had our own way of seeing things and our own concept of honor . . . many years had to pass so that I would begin to comprehend that our behaviour in Morocco . . . was the same as the Italians in Abyssinia. A mentality that has been educated by secular institutions such as the Army, the Church, etc. cannot be thrown overboard easily.[60]

The result was the formalization of a petition for the use of mustard gas on the 20th of August 1921;[61] however, as seems the case with Churchill, the wish for the immediate deployment of chemicals was not feasible at that time. Indeed, it was not until July 1923 that mustard gas artillery shells were fired for the first time on the Moroccans.[62] Such uses appear to have been restricted in part because of the limited supplies available, but also because of concerns over inflicting casualties on Spanish troops. Any concerns over harming Spanish troops were to some extent resolved by the summer of 1924 when mustard gas aircraft bombs were first used. This was followed by a period up until early 1926[63] in which significant quantities of chemical weapons, specifically mustard gas, were rained down on the people of the Rif, including on civilian populations in 'villages and crowded markets'.[64]

By the end of the war, Balfour suggested that gas weapons had killed hundreds of Rifians and severely injured thousands.[65] CW appear to have served a number of utilities over the course of the conflict. Spiers suggests that such weapons 'weakened the resistance of Abdel Krim's supporters',[66] in addition to which they appear to have been used to frighten indigenous Moroccans and seemingly to mete out the 'harshest punishment that must be felt with the greatest intensity'.[67]

ITALY IN THE SECOND ITALO-ABYSSINIAN CONFLICT

The third micro-case study concerns the Italian use of chemical weapons against Ethiopian tribes over the course of the Second Italo-Abyssinian conflict between 1935 and 1936. Italy had long had a fractious relationship with Ethiopia, having suffered a bruising defeat at the hands of Ethiopian Emperor Menilek's armies in the Battle of Adwa in March 1896. During the rise of what Mussolini referred to as the Third Rome in the 1930s, Ethiopia once again became a target of Italian designs, representing as it did an easy target for a country that Duggan describes as experiencing a powerful sense of 'asserting itself in the world, avenging past insults to the nation's dignity, and wiping away the image of weakness that liberalism had saddled it with'.[68] To this end, Fascist Italy deliberately sought a military solution to its claims on Ethiopia and over the course of 1934 and 1935 prepared the Italian population for the invasion through a concerted propaganda campaign targeted at Italians and designed to animate nationalist and xenophobic sentiments.[69]

Such was the expectation that Italy would swiftly and resolutely defeat the Ethiopian forces[70] that Mussolini required nothing less than a 'crushing victory'.[71] However, the 1935 campaign initially led by General Emilio De Bono begun to falter in the face of superior numbers of Ethiopian troops.[72] By mid-December and following the Ethiopian's Christmas Offensive,[73] Mussolini became increasingly worried that the promised victory would not be so easily achieved. Accordingly, he authorized Marshal Badoglio, who replaced De Bono, to use gas as required, including on a much larger scale. According to Duggan:

> Badoglio had in fact already begun employing mustard gas (or 'insecticide' as he euphemistically liked to call it) . . . and during the next three months around 1,000 heavy bombs filled with the chemical were dropped on enemy positions or, more lethally sprayed as vapour from aircraft, killing combatants and non-combatants alike and poisoning rivers and lakes.[74]

General Badoglio later argued the initial use of chemicals was experimental;[75] however, others have suggested this was a panicked reaction to an Ethiopian incursion. Whatever the initial reasons for turning to chemical weapons, over time this became considerably less experimental, with one estimate suggesting 15,000 casualties were caused by the use of gas in Ethiopia, with such weapons serving variously to protect Italian flanks, but also disrupt communications and demoralize Ethiopian forces.[76]

The use of chemical weapons was also facilitated by a deliberate and effective policy of othering on the part of the Italians. Such a policy is starkly apparent in the ensuing legislation applied – and rigorously enforced – after the conquest of Ethiopia[77] but also in the earlier engineering of clear ingroup and outgroup identities, something explicitly recognized by Mussolini who 'quite clearly explained to the Italian people that his anti African policies were meant to solidify Italian racial consciousness and solidarity'.[78] The process of othering and zoomorphistic dehumanizing of the Ethiopians appears to have had some success at least among some. References from the time analogize Ethiopians as 'bugs' requiring 'disinfection',[79] whereas as a collective of Bologna university students demanded Mussolini engage in a 'war to the limits . . . [against the] . . . inhuman, vile . . . bestial Abyssinian people . . . Chemical weapons should be used on a large scale as modern war demands'.[80]

As was the case with the Spanish in the Rif War, many of the perpetrators of chemical warfare appear to have privileged military and religious moral codes above any form of normative opprobrium against the use of such weapons. Certainly Duggan has stated that most soldiers 'were safely corralled inside a value system which patriotism, devotion to the Duce and obedience to the fascist regime intertwined with an exaltations of military virtues and a disparagement of liberal humanitarianism to blunt expressions of doubt or dissent'.[81] Such an approach was reinforced by notions that Italy was engaged in a civilizing mission, the details of which – including the use of chemical weapons – were not widely known within Italy where a policy of censorship[82] and denial[83] obfuscated the reality of the conflict for many Italians.

JAPAN IN MANCHURIA 1937

The fourth micro-case study concerns the use of chemical weapons by Imperial Japan against the Chinese, something covered in greater depth elsewhere in this volume by Jeanne Guillemin. This chapter does not seek to duplicate Guillemin's work but instead focuses on events in Manchuria through the lens of othering.

As early as the 1920s, Imperial Japan initiated small-scale experiments with gas,[84] something accelerated from 1933 onwards.[85] Following the Marco Polo Bridge incident,[86] Japan dispatched chemical weapons units to Manchuria, where they remained until the conclusion of the war in May 1945,[87] by which point, SIPRI has suggested, gas was 'sufficiently accepted by the military establishment to permit their development to an operational level'.[88] The use of gas by the Japanese was nonetheless initially cautious,[89] a trend that is perhaps consistent with Japanese guidance for use of gas warfare that emphasize careful investigation of the preparedness of the enemy for gas warfare and consideration of environmental factors.[90] However, as Spiers notes that as the conflict evolved, the 'increasing recourse to, and more diverse use of, poison gas indicates that some field commanders appreciated the utility of gas as a means of exploiting Chinese vulnerabilities'.[91]

Chinese sources have estimated a total of 889 chemical attacks were launched at military and civilian targets over the course of the conflict.[92] In terms of utilities, according to the translation of the Japanese document, *The Methods of Use of Gas*, 'the objective of the use of gas lies in causing enemy casualties, both personnel and horses, harassing their combat actions and eventually, in coordination with other arms of combat to bring about victory in the battle'.[93] The document goes on to provide advice on different uses of gas, including a passage which states that 'as soon as we perceive signs of retreat in the enemy we will drop non-persistent gas bombs on those areas where the enemy gathers or on the important points of passage on the route of retreat so that the enemy will be thrown into terror and confusion'.[94]

This was facilitated by the fact that the Chinese were seen as easy targets for gas warfare,[95] with translations of guidance for Chinese troops – as translated by U.S. officials – explicitly ordering troops not to 'make an unnecessary fuss about it but hold calmly to your position and receive the attack'.[96] However, the use of gas was also likely to have been aided by a powerful process of othering and extreme racism among the Japanese,[97] resulting in a situation in which racial superiority motivated Japanese activities and 'imparted an especially brutal character in Japanese imperialism, unique even among the imperialist powers'.[98] As much is supported by the records of U.S. indictments against Japanese defendants in the International Military Tribunal for the Far East (IMTFE), which stated that 'the mind of the Japanese people was systematically poisoned with harmful ideas of the alleged racial superiority of Japan over other peoples of Asia'.[99]

REFLECTIONS ON THE UGLY INTERWAR HISTORY

This chapter has shown that rather than just causing enemy casualties, chemical weapons have been employed by states for the purposes of, inter alia, harassing enemies, causing confusion, terrorizing combatants, punishing rebels and demoralizing otherwise-difficult-to-reach insurgent groups and sympathetic populations. Such utilities notably relied less on strategic weapons with gigantic effect and of the types and quantities envisaged by the superpowers during the Cold War[100] but smaller chemical weapons that exploited the ill-equipped and ill-prepared nature of populations. Based on these case studies, which are by no means representative of the history of chemical warfare, such weapons appear to be more likely in two particular circumstances: first, in conditions where an existential threat is faced and conventional forces are frustrated with the ineffectiveness of conventional means of attack; and second, in conditions of othering.

Concerning the first point, chemical weapons appear to have become more appealing following strategic shocks that threaten regime survival and following a period of frustration with the failure of conventional means and methods of war fighting against an unconventional enemy. In the case of the Tambov, the speed at which the peasant revolt spread was seen as the most significant threat to Lenin's nascent regime. Moreover, when it became apparent that conventional approaches to dealing with this threat were proving insufficient, chemical weapons were sought as an alternative to smoke out rebels from otherwise-hard-to-access areas. In the case of the Rif War, it was the Annual disaster that (re-)awakened interest in chemical warfare and lowered any moral barriers against their use. While Italy was hardly facing an existential threat, the barrage of propaganda that Italians were subject to meant that expectations were high and nothing short of a decisive Fascist victory would suffice. Yet contrary to such inflated expectations, Italy struggled to subdue the Ethiopians, and following the Christmas Offensive, Italy more brazenly turned to the use of alternative weapons. The Japanese began small-scale gas experiments in the late 1920s;[101] however, when the so-called China incident erupted in July 1937, Japanese gas units were sent to Manchuria and other provinces where they operated throughout the war until 8 May 1945.[102]

This is not a trend that is unique to the interwar years. Chemical weapons have since been used by Egypt in the North Yemeni Civil War, by South African and Rhodesian forces and by Saddam's regime in Iraq and Iran. In most cases, such weapons were used in relatively small quantities (at least compared to either the quantities of agent envisaged for use in strategic CW during the Cold War, or indeed the other forms of force used in these conflicts) by more powerful forces against ill-prepared and ill-equipped enemies in order to kill and diminish perceived threats to the respective regimes.

The second factor is that chemical weapons use by the powerful against the less powerful in the interwar years took place in the context of pervasive processes of othering that rendered the targets somehow inferior in the eyes of the chemical attackers. We argue this was important to overcome the cognitive dissonance generated by the use of weapons otherwise perceived as cruel or somehow taboo. This was realized through engineered narratives that dehumanized outgroups in various forms: in the Tambov rebellion, insurgents were portrayed as Class enemies, *Kulaks*, bandits and collectively as an 'infection' which required a 'shock campaign . . . to annihilate the disease'. In the Rif War, the forces of Abdel-Krim were seen as 'malicious beasts' and treacherous tribes that needed to be exterminated. In the case of Abyssinia, the Ethiopian tribes were variously analogized as 'bugs' that require destruction, or 'inhuman, vile . . . bestial Abyssinian people', in whose destruction 'any means were justified'. In the case of Japan, 'racism imparted an especially brutal character in Japanese imperialism, unique even among the imperialist powers'.[103]

This is not to suggest that the process of othering necessarily always leads to the use of chemical weapons. Moreover, there are cases, such as the use of chemical weapons in the North Yemeni civil war, where 'othering' was seemingly less vitriolic (although was still present).[104] Nonetheless, a process of othering and the dehumanization of the outgroup appears to help pave the way for the use of unconventional 'taboo' weapons by minimizing cognitive dissonance, something that can more readily be achieved in circumstances where an ingroup has suffered heavy casualties and experienced frustration with the existing means and methods of warfare available.

Such circumstances can also lead to a change in the norms that might otherwise inform and influence the perceived ingroup, or, more specifically, a slide in the moral baselines of the ingroup which potentially renders otherwise-unacceptable weapons acceptable, even banal. This is perhaps considerably easier in circumstances where states maintained greater control over the domestic media and were better able to both insulate domestic populations from the horrors of external conflicts and combatants from external scrutiny. In the Tambov, asphyxiating gas became merely one of a number of exotic weapons used in the cruel repression of the Antonov movement. In the Rif, the humanitarian framing of chemical weapons perhaps facilitated the shifting of moral baselines in a brutal conflict. Similarly, in Abyssinia, in cases where the use of chemical weapons was acknowledged, it was largely portrayed as a harmless incapacitant that was used sparingly in retaliation for acts of Ethiopian brutality;[105] and, as the Japanese case shows, in time chemical weapons became accepted by the military.

Since the interwar years, advances in information and communications technology have rendered the concealing of cruel acts comparatively more difficult. However, the same technology has also created opportunities for states to manipulate the narratives surrounding such events, including the apportioning of blame.[106] More significantly, a more robust regime preventing chemical weapons has been developed, specifically through the 1993 Chemical Weapons Convention (CWC), which, notably, makes no explicit distinction between civilized and uncivilized actors, thereby eliminating explicit language that could be exploited to legitimize 'othering' discourses. Moreover, this regime includes a robust inspection process designed to both 'serve as a confidence-building measure that each and every State Party is meeting its obligations under the Convention, and . . . to have a deterrent effect on those having the intention to contravene the Convention's provisions'.[107]

The world is better off for the CWC and those that tend to it; however, events, such as the use of chemical weapons in the Ghouta agricultural belt just outside Damascus, Syria, in August 2013, perhaps serve as a telling reminder that in circumstances where regimes perceive an existential threat from an ill-prepared, dehumanized 'outgroup' that cannot be readily dealt with by conventional means, chemical weapons retain some utilities in the fog of war as tools of the powerful against the weak.

NOTES

1 Richard M. Price, *The Chemical Weapons Taboo* (Ithaca, NY: Cornell University Press, 1997), 35.
2 See, for example: SIPRI, *The Problem of Chemical and Biological Warfare Volume I. The Rise of CB Weapons* (New York: Humanities Press, 1971).
3 Jean Pascal Zanders, 'International Norms against Chemical and Biological Warfare: An Ambiguous Legacy', *Journal of Conflict and Security Law*, 8, no. 2 (2003), 391–410.
4 Price, *The Chemical Weapons Taboo*, 26.
5 International Military Commission. *Declaration Renouncing the Use, in Time of War, of Explosive Projectiles Under 400 Grammes Weight* (Saint Petersburg, 1868).
6 Price, *The Chemical Weapons Taboo*, 35.
7 Brian Rappert and Alex Mankoo, 'Transgressive Chemicals', in *Chemical Bodies: The Techno-Politics of Control*, eds. Alex Mankoo and Brian Rappert (London: Rowman & Littlefield International Ltd., 2018).
8 Zanders, 'International Norms against Chemical and Biological Warfare'.
9 Price, *The Chemical Weapons Taboo*, 107.
10 War Office, *Manual of Military Law* (London: War Office, 1914), 235.
11 'Churchill's 1919 War Office Memorandum', National Churchill Museum, n.d., available at https://www.nationalchurchillmuseum.org/churchills-1919-war-office-memorandum.html.
12 Sebastian Balfour, 'The Secret History of Chemical Warfare against Moroccans', in *Deadly Embrace – Morocco and the Road to the Spanish Civil War* (Oxford and New York: Oxford University Press, 2002), 123–156; See also Saad Eskander, 'Southern Kurdistan under Britain's Mesopotamian Mandate: From Separation to Incorporation, 1920–23', *Middle Eastern Studies*, 37, no. 2 (2001), 176.
13 R. M. Douglas, 'Did Britain Use Chemical Weapons in Mandatory Iraq?', *The Journal of Modern History*, 81 (2009), 859–887.
14 Edward M. Spiers, *A History of Chemical and Biological Weapons* (London: Reaktion, 2010), 72.
15 Mike Crang, *Cultural Geography* (London: Routledge, 1998), 61.
16 Henri Tajfel, 'Social Psychology of Intergroup Relations', *Annual Review of Psychology*, 33, no. 1 (1982), 1–39.
17 M. A. Hogg and Dominic Abrams, 'Towards a Single-Process Uncertainty-Reduction Model of Social Motivation in Groups', in *Group Motivation: Social Psychological Perspectives*, eds. M. Hogg and D. Abrams (London: Harvester Wheatsheaf, 1993), 173–190.

18 M. B. Brewer, 'Identity and Conflict', in *Intergroup Conflicts and Their Resolution: A Social Psychological Perspective*, ed. Daniel. Bar-Tal (London: Psychology Press, 2011), 125–143.
19 Waldemar Lilli and Jürgen Rehm, 'Judgmental Processes as Bases of Intergroup Conflict', in *The Social Psychology of Intergroup Conflict*, eds. W. Stroebe, A. W. Kruglanski, D. Bar-Tal and M. Hewstone (New York: Springer-Verlag Publishing, 1988).
20 Terrell A. Northrup, 'The Dynamic of Identity in Personal and Social Conflict', in *Intractable Conflicts and Their Transformation*, eds. Louis Kriesberg, Terrell A. Northrup and Stuart J. Thorson (New York: Syracuse University Press, 1989).
21 George A. Kelly, *A Theory of Personality: Psychology of Personal Constructs* (New York: Norton, 2013).
22 Brewer, 'Identity and Conflict,' 125–143.
23 Ibid.
24 Albert Bandura, 'Moral Disengagement in the Perpetration of Inhumanities', *Personality and Social Psychology Review*, 3, no. 3 (August, 1999), 193–209.
25 Daniel Bar-Tal and Phillip L. Hammack Jr., 'Conflict, Delegitimization, and Violence', in *The Oxford Handbook of Intergroup Conflict*, ed. Linda R. Tropp (Oxford: Oxford University Press, 2012).
26 Daniel Bar-Tal, *Intractable Conflicts: Socio-Psychological Foundations and Dynamics* (Cambridge: Cambridge University Press, 2013).
27 Bandura, 'Moral Disengagement in the Perpetration of Inhumanities'.
28 Stéphane Courtois, Nicolas Werth, Jean-Louis Panné, Andrzej Paczkowski, Karel Bartošek and Jean-Louis Margolin. *The Black Book of Communism: Crimes, Terror, Repression* (Cambridge, MA: Harvard University Press, 1999), 109.
29 Orlando Figes, *A People's Tragedy: The Russian Revolution 1891–1924* (London: The Bodley Head, Vintage, 2017).
30 Murray Bookchin, *The Third Revolution: Popular Movements in the Revolutionary Era* (London: Cassell, 1996).
31 Richard Pipes, *Russia under the Bolshevik Regime* (New York: Vintage Books, 1995), 386.
32 Sheila Fitzpatrick, 'Ascribing Class: The Construction of Social Identity in Soviet Russia', *The Journal of Modern History*, 65, no. 4 (December 1993), 745–770.
33 Alexander Statiev, *The Soviet Counterinsurgency in the Western Borderlands* (Cambridge: Cambridge University Press, 2010).
34 Ibid., 14.
35 Ibid., 16.
36 Courtois et al., *The Black Book of Communism*, 116.
37 Pipes, *Russia under the Bolshevik Regime*, 386.
38 Erik C. Landis, *Bandits and Partisans: The Antonov Movement in the Russian Civil War* (Pittsburgh: University of Pittsburgh, 2008), 265.
39 Courtois et al., *The Black Book of Communism*, 116.
40 Landis, *Bandits and Partisans*, 266.
41 Alexander S. Bobkov, 'On the Issue of Using Asphyxiating Gas in the Suppression of the Tambov Uprising', *Journal of Slavic Military Studies*, 25, no. 1 (2012), 65–104.

42 Landis, *Bandits and* Partisans, 66.
43 Bobkov, 'On the Issue of Using Asphyxiating Gas in the Suppression of the Tambov Uprising'.
44 Landis, *Bandits and Partisans*, 265.
45 Bobkov, 'On the Issue of Using Asphyxiating Gas in the Suppression of the Tambov Uprising'.
46 Ibid.
47 Ibid., 71.
48 Courtois et al., *The Black Book of Communism*, 116.
49 Bobkov, 'On the Issue of Using Asphyxiating Gas in the Suppression of the Tambov Uprising'.
50 Spiers, *A History of Chemical and Biological Weapons*, 72.
51 Balfour, 'The Secret History of Chemical Warfare against Moroccans', 131.
52 Ibid.
53 Crespo de Lara, 'Inocencia O Errores Fatales', *La Correspondencia Militar*, 10 October 1921, available at http://hemerotecadigital.bne.es/issue.vm?id=0002178083&search=&lang=es.
54 As translated from Juan Pando, *Historia secreta de Annual*, in which the original reads 'Siempre fui refractario al empleo de los gases asfixiantes contra estos indígenas, pero después de lo que han hecho, y de su traidora y falaz conducta, he de emplearlos con verdadera fruicción'. Juan Pando, *Historia Secreta de Annual* (Madrid: Ediciones Temas de Hoy, 1999).
55 Balfour, 'The Secret History of Chemical Warfare against Moroccans', 135.
56 Ibid.
57 Pando, *Historia Secreta de Annual*.
58 Balfour, 'The Secret History of Chemical Warfare against Moroccans', 137.
59 Ibid., 137–138.
60 Ignacio Hidalgo de Cisneros, *Memorias: Cambio de Rumbo: Volume 1* (Paris: Société d'Éditions de la Librarie du Globe, 1964).
61 Pando, *Historia Secreta de Annual*.
62 Raimon Guitart, *Tóxicos: Los Enemigos de La Vida* (Barcelona: Edicions UAB, 2014).
63 Balfour, 'The Secret History of Chemical Warfare against Moroccans', 128.
64 Ibid., 152.
65 Ibid., 153.
66 Spiers, *A History of Chemical and Biological Weapons*, 73.
67 Balfour, 'The Secret History of Chemical Warfare against Moroccans', 143.
68 Christopher Duggan, *Fascist Voices: An Intimate History of Mussolini's Italy* (London: The Bodley Head, 2012), 255.
69 Philip Morgan, *Italian Fascism, 1915–1945*, 2nd ed. (London: Palgrave MacMillan, 2004), 173.
70 Ibid.
71 Duggan, *Fascist Voices*, 257.
72 Ibid., 261.
73 Lina Grip and John Hart, 'The Use of Chemical Weapons in the 1935–36 Italo-Ethiopian War', *SIPRI Arms Control and Non-Proliferation Programme*

(October 2009), 22–25 , available at https://www.sipri.org/sites/default/files/Italo-Ethiopian-war.pdf.
74 Duggan, *Fascist Voices*, 261.
75 N. J. McCamley, *The Secret History of Chemical Warfare* (Barnsley, Yorkshire: Pen & Sword Books, Ltd, 2006), 45.
76 SIPRI, *The Problem of Chemical and Biological Warfare Volume I*, 142.
77 As Gillette notes: 'on August 5, 1936, only several months after the conquest of Ethiopia, a decree was issued outlining the system of racial separation that would now characterize Italian East Africa. Beginning with the fundamental provision that "the lives of whites and blacks should be completely separate in Italian East Africa". . . the Police were instructed to enforce these provisions with "extreme rigor"' Aaron Gillette, *Racial Theories in Fascist Italy* (London and New York: Routledge, 2002), 57.
78 Gillette, *Racial Theories in Fascist Italy*, 58.
79 R. J. B. Bosworth, *Mussolini's Italy: Life under the Dictatorship, 1915–1945* (New York: Penguin Books, 2006).
80 Duggan, *Fascist Voices*, 254.
81 Ibid., 264.
82 Ibid., 262.
83 SIPRI, *The Problem of Chemical and Biological Warfare Volume I.*
84 Spiers, *A History of Chemical and Biological Weapons*, 76.
85 *Intelligence Report on Japanese Chemical Warfare,* Vol I (Tokyo: Office of the Chief Chemical Officer, GHQ, 15 May 1946), available at Combined Arms Research Digital Library, https:/server16040/contentdm.oclc.org.
86 James W. Hammond, *Poison Gas: The Myths versus Reality* (Westpoint, CT:Greenwood Press, 1999).
87 Spiers, *A History of Chemical and Biological Weapons*, 76.
88 SIPRI, *The Problem of Chemical and Biological Warfare Volume I.*
89 Spiers, *A History of Chemical and Biological Weapons*, 76.
90 IN: Office of the Chief Chemical Officer, Intelligence Report on Japanese Chemical Warfare, vol 1 (1946) http://bulletpicker.com/pdf/Intelligence%20Report%20on%20Japanese%20Chemical%20Warfare,%20Vol%20I.pdf.
91 Spiers, *A History of Chemical and Biological Weapons*, 77.
92 McCamley, *The Secret History of Chemical Warfare*, 45.
93 *Intelligence Report on Japanese Chemical Warfare*, Vol I, 127–128.
94 Ibid., 133.
95 Ibid., 22.
96 Office of the Chief Chemical Officer, 'Chapter VI – The Chinese Army as Seen from the Point of View of Chemical Warfare', 124.
97 Louise Young, *Japan's Total Empire: Manchuria and the Culture of Wartime Imperialism* (California: University of California Press, 1999), 370.
98 Tsuyoshi Hasegawa and Kazuhiko Togo, *East Asia's Haunted Present: Historical Memories and the Resurgence of Nationalism* (London: Praeger Security International, 2008).
99 Neil Boister and Robert Cryer, eds., *Documents on the Tokyo International Military Tribunal: Charter, Indictment and Judgments* (Oxford: Oxford University Press, 2008).

100 Alex Ghionis, Implications for the Chemical Weapons Convention after Syria, (Sussex: MA dis, University of Sussex, 2014), 24.
101 Spiers, *A History of Chemical and Biological Weapons*, 76.
102 Ibid.
103 Hasegawa and Togo, *East Asia's Haunted Present*.
104 Asher Orkaby, *Beyond the Arab Cold War: The International History of the Yemen Civil War, 1962–68* (Oxford: Oxford University Press, 2017), 33.
105 Morgan, *Italian Fascism, 1915–1945*, 262.
106 James Revill, Alex Ghionis, Caitríona McLeish and Steve Johnson, 'Ghouta Narratives: A Critical Assessment of Syrian Chemical Weapons Narratives', Harvard Sussex Program Syria Collection, 2016.
107 Organisation for the Prohibition of Chemical Weapons, 'The Verification Process', 2017, available at https://www.opcw.org/our-work/non-proliferation/the-verification-process/.

Chapter 4

Biological Warfare, Chemical Warfare and the Public Body

Etienne Aucouturier

It is commonly accepted that biological and chemical warfare denote the uses of two different types of weapon systems. Simply put, on the one hand, chemical warfare is the use, in armed conflict, of inert, synthetic, chemical substances, in order to kill or incapacitate enemies. On the other, biological warfare is the use of living organisms or of products of living organisms (such as toxins) in armed conflicts in order to kill or incapacitate enemies.

This chapter asks whether this distinction between chemical and biological weapons might be challenged. I argue that both weapon systems make sense only with regard to their intended use and should, for theoretical, legal and ethical reasons, be reconnected under one umbrella concept. Although both systems can be used for similar purposes, it is also commonly accepted and legally acknowledged that these systems do indeed differ in some important respects: for one, chemical warfare agents diverge from some biological warfare agents in not sharing the property of being contagious, as well, of course, as in that of not being alive. Some inert biological agents (mainly toxins such as ricin), however, do not differ significantly, as far as their operational purpose is concerned, from synthetic toxins: these could just as well be termed 'organic' chemical weapons, without major theoretical loss. Furthermore, the targets of both chemical and biological weapons systems can be either national populations or natural resources. Selecting targets involves the political act of sorting bodies to be protected from those to be attacked – for example, through preventative medical countermeasures such as gas masks or vaccines. Thus, chemical and biological warfare also constitute forms of physiological warfare that can target a population (of humans, animals or plants) in a given environment or political context without harming another, and in this sense entail forms of political ordering.

In this chapter, I shall generally challenge the classical delimitations of biological and chemical warfare. Many characteristics of both chemical and biological weaponsystems, as well as their joint history, strongly suggest a potential reconnection under one umbrella concept. I argue that this concept is preferably biological warfare, or a more target-based notion of *physiological warfare*, where the target is physiological functions in living systems, not inert matter: nerve gases do not do any harm to buildings, but to nervous systems, for instance. On the basis of existing definitions and legal frameworks, I mainly ground the reconnection of biological warfare with chemical warfare on philosophical, bioethical and epidemiological arguments.

The first three sections highlight and comment on the existing definitions of biological and chemical warfare and their variations throughout history, as well as the legal frameworks governing these specific forms of war. Then, the chapter evaluates these definitions, their objects and relevance through posing a number of questions: Do these technical definitions and designations miss the point that biological warfare and chemical warfare are both in principle political and in teleological (goal-oriented) notions, rather than actual biological or chemical notions? Given that physiological targets define the means of biological warfare and chemical warfare (not the reverse), should not the intended effect on a target be the common notion to define and morally (then legally) evaluate these means? Finally, if we understand the epidemiology of biological warfare and chemical warfare as the measure of occurrence of diseases intentionally produced (or inadvertently provoked), is it still relevant, and practical or beneficial, to maintain an artificial distinction between means to target selectively physiology, from a global health point of view?

EARLY DEFINITIONS

Since the second half of the nineteenth century, several conferences, codes or conventions mentioned the prohibition of the use of toxic or poisonous weapons at war.[1] The 1874 Brussels Conference on the Laws and Conventions at War prohibited the use of 'poison or poisonous weapons'; delegates at the two following international peace conferences at the Hague (1899 and 1907) agreed to ban 'the diffusion of poisonous or deleterious gases'; and the 1925 Geneva Protocol (GP) prohibited both 'the use in war of asphyxiating, poisonous or other gases, and of bacteriological methods of warfare'. Although this seemed to, for the first time, clearly legally differentiate between chemical and bacteriological warfare, it is worth noting that, in fact, the objective of the GP was to ban the wartime use of microscopic agents able to provoke diseases. The broad sense of 'poison' or 'poisonous' was clearly referring to the effects produced on potential targets of a variety of substances. Under these laws and regulations, the action of poisoning during wartime became forbidden, whatever the poison.

Until 1945, and particularly after the GP, the use of bacteriological poisoning became alternatively named 'microbial' warfare, 'bacterial' warfare or 'bacteriological' warfare. These denominations were in line with the contemporary germ theory, which was yet to include viruses, prior to their identification in the 1950s. These substantial definitions, based on the types of poisons used to wage war, were rather coherent with the designated agents and introduced a distinction between toxic poisons and contagious ones.

In 1945, a more comprehensive definition of biological warfare was formulated by George W. Merck (then advisor to the Secretary of War of the United States), describing the agents of biological warfare, including viruses, as well as their intended targets (men, animal and plant physiology):

> Biological warfare may be defined as the use of bacteria, fungi, viruses [rickettsia] and toxic agents from living organisms (as distinguished from synthetic chemicals used as gases or poison) to produce death or disease in men, animals, or plants.[2]

We shall discuss this distinction based on the 'synthetic' character of toxic agents given further. This definition was later, in 1950, adopted by the French military, who considered this denomination more accurate than the then most used 'bacterial' (French 'bactérienne') warfare.[3] Physician and colonel Jude pointed out that this more inclusive notion was, by including non-living biological agents (toxins), bringing biological weapons closer to chemical weapons.[4] An outsider concept also emerged in 1962, with sociologist Gaston Bouthoul's 'antibiotic weapons'.[5] Although he seemed to identify a new category of weapons of mass destruction (WMD) selectively targeting living beings, he did not include in it biological weapons:

> Antibiotic weapons create new destructive possibilities whose characteristic is that they allow to eliminate all human life in a determined area. In this way, asphyxiating gases and vesicants, radioactive gases and clouds, and nuclear explosives producing altogether explosive and antibiotic effects such as the current uranium atomic bomb, which seems to be destined to be soon replaced by the hydrogen bomb and the cobalt bomb.[6]

Although a bit confusing, this explanation made sense of the dissociation of this 'antibiotic' effect from other mechanical or explosive effects, technically reached with the development of modern chemical and biological warfare techniques. This particular effect was sought as the norm of both weapon systems.

The histories of national chemical and biological warfare programmes reveal a very close intertwining of institutions, scientists, techniques, places, facilities, laboratories and testing grounds dedicated to both categories of weapons.[7] The French endeavours to produce chemical and biological weapons and warfare programmes make no distinction between the two: scientists, technicians and military personnel were constantly working on both weapon systems at the same time and the same places from the decade before the World War I.[8] The main first leading figure of the French chemical and biological warfare programme, Auguste Trillat, was, for instance, both a chemist and epidemiologist at the Institut Pasteur, applying his knowledge and techniques of disinfection through chemical means (formaldehyde) to work on defence against biological warfare.[9] He also specialized in producing artificial clouds, which could convey synthetic toxins or biological agents. Furthermore, later climatology and studies on the chemical and physical conditions of the atmosphere were pursued, in order to improve the dispersion of biological warfare agents in the battlefield.[10] Chemical and biological warfare nevertheless became gradually judicially separated in differing sets of laws (see the next section), while the convergence of the two categories was sought in contexts of both offensive and defensive military research, alongside chemical, biological and nuclear weapons gradually being considered together as WMD.

Confusion persisted (and still exists) on the differentiation between chemical and biological weapons, and between biological weapons themselves. The very definition of the object prohibited by the 1972 Biological and Toxin Weapon Convention (see the paragraph below on legal issues), the only global treaty prohibiting biological warfare, is still ambiguous: its title itself names the related weapons as 'bacteriological (biological)'. It is surprising that such an international treaty did not mention more accurately the very object of its prohibition.

CURRENT DEFINITIONS

Chemical and biological warfare are currently defined separately in military classifications, which determine their legal frameworks, and military deployments. In the current definition of reference for all North Atlantic Treaty Organization members, in the Department of Defense (DOD) Dictionary of Military and Associated Terms,[11] biological warfare and chemical warfare are defined as follows:

- Biological warfare: 'Employment of biological agents to produce casualties in personnel or animals, or damage to plants or materiel; or defence against such employment'.[12]
- Chemical warfare: 'All aspects of military operations involving the employment of lethal and incapacitating munitions/agents and the warning and protective measures associated with such offensive operations. Also called CW. See also chemical agent; chemical weapon; riot control agent'.[13] In the same dictionary, the related weapon systems are so defined:
- Biological weapon: 'An item of materiel which projects, disperses, or disseminates a biological agent including arthropod vectors'.[14]
- Chemical weapon: 'Together or separately, (a) a toxic chemical and its precursors, except when intended for a purpose not prohibited under the Chemical Weapons Convention; (b) a munition or device, specifically designed to cause death or other harm through toxic properties of those

chemicals specified in (a), above, which would be released as a result of the employment of such munition or device; (c) any equipment specifically designed for use directly in connection with the employment of munitions or devices specified in (b), above. See also chemical agent; chemical warfare; riot control agent'.[15] The concomitant agents are so defined:

- Biological agent: 'A microorganism (or a toxin derived from it) that causes disease in personnel, plants, or animals or causes the deterioration of materiel. See also chemical agent'.[16]
- Chemical agent: 'A chemical substance that is intended for use in military operations to kill, seriously injure, or incapacitate mainly through its physiological effects. See also chemical warfare; riot control agent'.[17]

These technical definitions include goals, purposes or intentions, some of them common to both weapon systems, as well as wide scopes of potential contexts of use. These goal-oriented notions revolve around toxic or pathogenic effects of both weapon systems (while by contrast, harmless germs or substances are irrelevant to biological or chemical warfare): what is sought and central through both practices is the intention to produce epidemics (that are not necessarily contagious). I use the term 'epidemic' here to refer a widespread occurrence of a physiologically harmful phenomenon.

However, in human, animal or plant populations, organisms generally share many natural physiological and epidemiological vulnerabilities, such that external defence systems or countermeasures, like gas masks or other protective devices (in the case of humans and animals), or internal medical or immunological ones, are required to protect against these weapons. In this respect, the potential for chemical and biological weapons to affect the troops using them has been a concern for military institutions involved in CBW programmes. As such, biological and chemical warfare imply sorting through and identifying populations (humans, animals or plants) and their physiological vulnerabilities, before taking advantage of these vulnerabilities for hostile purposes. For example, a choking gas will harm an individual with a breathing condition more than someone without one, and someone without a gas mask more than someone with one.

In the DOD definition, biological warfare includes 'defence against' the employment of biological agents. Likewise, chemical warfare includes 'warning and protective measures associated with' operations involving chemical agents. In his controversial 1968 book on biological warfare, *We all fall down*,[18] Robin Clarke considered biological warfare as 'public health in reverse', meaning that public health institutions have to work simultaneously and secretly at understanding destructive effects by searching to improve the virulence and stability of pathogens. Clarke also advocated a designation of chemical and biological weapons within one category of weapons targeting the living in the form of an 'ecological warfare'. This point of view, linked to the consequences of the massive use of defoliants in the Vietnam War at the time, and later shared by others,[19] was severely criticized by the then Director of Porton Down's Microbial Research Establishment (UK), Dr Charles Edward Gordon Smith. Cited in Clarke's book, Smith acknowledged that the notion of 'public health in reverse' was quite accurate but rejected taking chemical and biological warfare to be one category on the following grounds: 'the essential difference between B and C agents is that only an infective dose of a B agent is required as it will multiply to become toxic, while a full toxic dose of C agents must be administered – bacterial toxins are thus C agents'.[20]

This distinction based on proliferation of an agent through the host is a crucial one when considering the possible reconnection of biological warfare and chemical warfare. But as Gordon Smith points himself out in his objection, 'Bacterial toxins are thus C agents'. I shall argue further that this proliferation limitation constitutes at best a legal or nomenclature limitation, but that it is not ethically and epidemiologically relevant: biological and chemical warfare are not processes of science but political actions that use technology from the biomedical sciences and chemistry to anticipate the effects of an artificially created epidemic and control particular bodies. Biological warfare does not in this sense compare to nuclear warfare and deterrence, in that it can include (provided that it is not completely suicidal) a selectively distributed remedy to a doomsday bug. Despite some post-war fantasized attempts to produce 'vaccines' against radioactivity,[21] nuclear deterrence does not include so far this selective bodily or immunological protection of individual members of a political community, and the effects of nuclear weapons are not only physiological or 'antibiotic', as Bouthoul put it, but also mechanical.

LEGAL ASPECTS

The use, production and stockpiling of biological and chemical weapons are currently prohibited and arguably regulated by a series of international treaties, legal texts and implementation measures. The main international treaties banning biological and chemical weapons are the Geneva Protocol (1925), the Biological and Toxin Weapons Convention (BTWC, 1972) and the Chemical Weapons Convention (CWC) (CWC, 1993). The two most recent conventions included a commitment on the part of the signatories to destroy existing stocks. Thirty-nine states have to this day not signed the BTWC or have yet to accede, and three have yet to accede or sign the CWC. Thus, while biological warfare was forbidden since the beginning of the 1970s, chemical warfare remained by default authorized until the mid-1990s. With this separation, biological and chemical warfare became, in the second half of the twentieth century, governed by differing sets of laws and implementation measures.

Implementation and verification measures by institutions supplement this general international legal framework. The CWC benefits from the activities of an international organization, the Organisation for the Prohibition of Chemical Weapons (OPCW),[22] which conducts verification measures and promotes international conformity to the convention. This is not the case with the BTWC, which to some extent explains the relative difficulty the BTWC has had with regard to implementation compared to the CWC.[23] Furthermore, as shown in the previous paragraph, if defensive or protective measures are to be included in the definition of biological warfare, it poses difficulties for the task of distinguishing military from civilian technology, and for verifying if activities are in conformity with the BTWC.[24]

The main political mechanism, within the BTWC meant to generate transparency regarding activities related to biological warfare in the sense of the BTWC, is the so-called *confidence building measures* (CBM). These were created in 1986, with a view to disambiguate research activities of signatory countries and promote international cooperation in these strategic research fields[25] (with data exchange on research, on outbreaks, incitation to publish results, promotion of contracts, declarations of national legislations, declaration of past offensive and defensive activities and declaration of vaccine production units). However, many limitations impede the efficiency and credibility of the CBM reports. They are confidential and no verification measures have been put in place alongside them. The primary verification mechanism of the past was the ad hoc group of governmental experts to identify and examine potential verification measures from a scientific and technical standpoint (VEREX). With limited powers, VEREX has gathered only four times, in 1992 and 1993. The credibility of VEREX has been so challenged that diplomats in the 1990s referred maliciously to it as the 'V-word'.[26]

The setbacks encountered in the implementation of the two sets of laws, by the institutions governing biological warfare and chemical warfare, respectively, could benefit from an integrated understanding of the very objects of the prohibitions, which is more the intention to target physiology, than biological or chemical substances themselves.

The major twenty-first-century international regulation of reference is Resolution 1540[27] (meeting n°4956) adopted by the UN Security Council on 28 April 2004. Resolution 1540 seeks to counter the 'proliferation of nuclear, chemical and biological weapons, as well as their means of delivery'. Among other concerns, the 2001 terrorist attacks on the United States (including the anthrax letters[28]) have motivated this resolution to reinforce the international monitoring and control of all activities in nuclear science, biology and chemistry with potential military applications. Non-state actors, that is, all 'individual[s] or entit[ies], not acting under the lawful authority of any State in conducting activities which come within the scope of this resolution',[29] and counterterrorism were at the centre of this new resolution, with a view to take into account and monitor potential non-state armed conflicts.

SCALE MATTERS: ON THE SEEMINGLY ANTI-REDUCTIONIST UNDERSTANDING OF BIOLOGICAL WARFARE

The difference between biological and chemical weapons on the one hand and other weapon systems on the other lies in the fact that they are not intended, or only marginally, to destroy or damage inert things. They share this main common property of being intended to destroy or incapacitate living beings or systems without destroying their inert surroundings. In this sense, they are intended to discriminate or selectively target the living, or sub-systems of the living (e.g. nervous systems, the intended target of nerve gases). Indeed, chemical weaponry excludes military chemistry applied to other purposes: incendiary weapons or explosives, which are not intended to discriminate between the inert and the living, by targeting given physiological properties in a given environment, are not considered chemical weapons and are defined separately in the DOD Dictionary of Military and Associated Terms.

Biological weapons, being restricted to microorganisms or toxins derived from microorganisms, are also intended to target selectively the living or sub-systems of the living: although the current definitions of biological warfare and biological agents include damaging 'materiel' or the 'deterioration of materiel', the behavioural properties, if any, of biological agents or biological warfare are generally excluded, and no historical example exists to this day of a use of this particular weapon system against materiel (although research on the destruction of textiles by arthropods, e.g., has been of military interest). Arthropods (e.g. insects and acarids) are mentioned only as 'vectors', delivery means, of micro-biological pathogens and are coined 'biological' only in this sense, like inert rockets or cluster bombs designed to convey chemical or biological agents are.

In all, the current definition of biological warfare omits most biological organisms in conflict: Why does it exclude the use in armed conflicts of mechanical or cognitive properties, or functions, of other living beings of higher levels of complexity than microorganisms or molecules? Why wouldn't arthropods themselves also be considered as biological agents, not only as vectors? Arthropods have been used as weapons in diverse military operations for their ability to target crops selectively.[30] Many other living beings, including cattle and horses, have been also extensively used in warfare for the transportation of troops or mounted battle. Dolphins and other sea mammals have been trained to defend or detect ships, or to attack divers.[31] And after all, cognitive, psychological and mechanical properties of human beings themselves, trained or under the influence of a diversity of drugs,[32] are of course at the centre of any armed conflict. Humans can themselves intentionally and selectively target the living in a given environment. Since biology is not limited to microbiology or molecular biology, why exclude most biological entities from the category of *agent* of biological warfare? Is biological warfare only reductively qualified as 'biological', for practical, operational, political or legal purposes?

It seems obvious at this stage that the core selection criterion to qualify a biological weapon is scale, with the implication that a biological weapon is something that is invisible to the naked eye that produces visible symptoms or death. Chemical weapons share this feature.

If the core feature of these physiology-targeting weapons was the nature of agents used, biological warfare would be almost indefinable, as 'biological' can refer to any living agent able to selectively provoke damages to other living beings: a trained attack-dog, a human sniper or a mushroom, could altogether be considered biological weapons. And as biology does not only include microbiology, but also evolution and natural selection, one could even further argue that any form of war, from a naturalistic point of view, could qualify as 'biological', provided that it is a recurring behaviour of humans participating in natural selection. With humans being able to produce synthetic toxins, one could argue that products of chemical synthesis are 'derived from' an organism able to produce them, just like the biological toxins 'derived from' microorganisms are included in the DOD's definition of a biological agent. Of course, this is not what the military, politicians, scientists, physicians or legislators mean by biological warfare for instrumental purposes. What they mean is using poisons in war. So if one rejects this very wide conception of biological warfare, one has to acknowledge that scale matters more than the nature of the agents used, and specifically the scale of what is targeted, namely physiological functions. So-called chemical agents can, in the same way, be considered as chemical weapons only according to this function.

Intentionality, and the intention to produce physiological damage, might thus be retained as the central and most important feature in biological warfare, understood rather as physiological warfare: the physiological properties of the host (and to some extent the environment or the biophysical properties of biotopes) dictate the means of destroying or altering it. As far as the goal of biological warfare is concerned, it does not significantly matter if the same organ or physiological function is permanently or temporarily damaged or destroyed by a microorganism, a chemical molecule, a toxin produced by a living being or even by a nanotechnological device or a microscopic device with derived intentions,[33] such as a robot.

In seeking a convergent and synthetic definition of biological warfare including chemical warfare, I shall now suggest that the definition of biological warfare should include, at least, the employment of chemical agents, which share decisive properties with biological agents. I shall here adopt a preliminary modified definition of biological warfare to include the employment of weapon systems designed to selectively target the living, or sub-systems of the living, and them only, in the form of poisons invisible to the naked eye – this includes so-called chemical agents as defined earlier in the chapter.

Furthermore, artificially separating the use of microscopic or smaller entities to provoke physiological damages in humans, plants or animals from other means of doing so overlooks the fact that these means are constantly changing with the advancements of science and technology. Nanotechnology, as well as base editing techniques combining chemistry and molecular biology, is nowadays, for instance, challenging the very categories of the biological and the chemical. But even less spectacular and older disciplines, such as biochemistry, already blurred the demarcation between a strictly biological understanding of physiology and a chemical one. Hence the intention to target biological functions through the use of any microscopic or even nanoscopic invasive technique could be missed such that these uses do not fall under either category of biological or chemical warfare as currently defined, although they would in the same way produce selective damage on physiological functions in populations of living beings. The seemingly anti-reductionist separation between the biological and the chemical in these weapon's definitions ignores their core mode: targeting physiological functions.

This physiological criterion – demarcating what kinds of microscopic, or smaller, entities, can qualify as biological (or physiological weapons) – requires an understanding of what is harmful and what is not, or how microorganisms, substances or other invisible or inert devices can modify these functions. As stated earlier, any generally not harmful chemical substance (e.g. sugar, generally), microorganism (e.g. bacteria of the intestinal microbiota, generally) or microscopic device (e.g. implantable medical nanotechnology, generally) would not be considered a biological weapon or chemical weapon as currently defined. Else we would legally deprive ourselves from natural or technological kinds essential to our lives (e.g. our microbiota). Hence is it the expected pain or destructive effect on physiological functions in the targeted bodies that define a part of these natural or technical kinds as weapons.

BIOETHICAL INTERSECTIONS: ACTUAL SURGICAL STRIKES, QUALITY OF DEATH AND *MAUX SUPERFLUS*

What difference does it make to be the target of biological or chemical weapons? And why would these weapons be crueller than others? Biological and chemical warfare have historically featured very negatively in the press, in popular culture, and their mere evocation can be terrifying. The terror they instil is so great that hoaxes are both startling and frequent,[34] while actual casualties have been rather rare (and extremely rare in the case of so-called biological warfare) throughout the twentieth century and to the present day, particularly compared to other less-feared risks (e.g. gun-related casualties, which cause between 400,000 and 600,000 deaths per year worldwide[35]). The restriction in official definitions of biological warfare to the employment of biological agents that can produce a massive pandemic has also frequently been echoed in arts and popular culture, where biological warfare usually means the catastrophic and mass destructive use of only contagious, invisible, biological agents to kill human populations, such as in Jack London's *An Unprecedented Invasion* (1910),[36] and in many other films.[37]

Throughout the history of biological and chemical weapons programmes (as defined earlier), the military have generally considered their offensive use as unfair and unnecessarily cruel, while actively preparing defence and retaliation against enemy use. This has entailed two-way arms races, involving the production of physiological warfare agents potentially secretly produced by other parties, in order to prepare relevant vaccines, remedies and other protective measures.[38] Analysts, commentators and other authors also regularly depict biological and chemical weapons as particularly or excessively immoral. For example, journalists Robert Harris and Jeremy Paxman write in *A higher form of killing*:

> They are, first and foremost, indiscriminate weapons – 'dirty', as one young soldier we met during our researches put it. By and large they are invisible, and do their damage from within the body. You may not see the bomb or the bullet that kills you, but that external threat is somehow 'cleaner' than the malignant tumor, the paralysis or suffocation inflicted by these unseen weapons.[39]

This idea or moral intuition that CBW weapons insidiously damage the body from within, as actual surgical weapons, is also in tune with some military discourses about biological and chemical weapons. Inspired by Claude Bernard's physiology, as early as 1926, French military instructors compared chemical warfare agents to microscopic scalpels, able to target organs selectively in animal or human populations.[40] These moral intuitions are often formulated through, and grounded in, earlier tragic experiences, among which are the abundantly documented sufferings provoked during World War I gas attacks.

On the whole, in popular culture as well as in the military sphere, biological and chemical warfare are usually depicted as insidious, dreadful, highly uncontrollable, immoral and potentially catastrophic. Biological weapons are seen as very cruel WMD, with potential effects comparable to those of atomic weapons. While, in contrast, the use of incapacitating crowd control chemical weapons by police forces is still quite commonly accepted in peacetime in demonstrations worldwide, it is doubtful that the same acceptance would occur should police forces in these circumstances use non-lethal micro-organisms on civilian populations. It seems that the idea of invisible biological agents purposively used to cause a disease or disorder in a human population is even more intolerable to most, although other deadly practices are more easily accepted.

The social perception of this special type of risk, identified as mysterious, insidious, potentially catastrophic and unjustified, is in this author's view biased and missing the point that biological and chemical warfare are aimed at specifically targeting human physiology, unlike conventional weapons. It has been a constant concern of any military CBW programme to anticipate the backlashes of producing, and using in combat, dangerous biological or chemical substances. These programmes anticipate the pathogenic effects CBW might have on the enemy, and how dispersion of agents might be controlled. This has been a constant technical challenge; hence, a relative reluctance in the military sphere to use such weapons, and the multidisciplinary aspect of CBW programmes (which included, e.g., meteorological endeavours).

But some marginal apologists have also defended the use of biological and chemical weapons as opposed to conventional weapons. Population geneticist, Marxist and pacifist John Burdon Sanderson Haldane[41] defended the use of chemical weapons against the less predictable and deadlier use of conventional weapons after World War I in his provocative 1925 pamphlet *Callinicus*. In a style suffused with scientism, he argued that it was retrograde to ban chemical weapons (the terms of the GP were being discussed), while still allowing one 'to push or throw pieces of metal at the enemy',[42] only on the grounds that they were considered 'unusual and unpleasant'.[43] After testing many chemical agents on himself during the war and making a typology of their physiological effects, he argued that, if war should be regulated, it should not be on militaristic arguments such as the unsoldierly character of invisible gases but rather on the knowledge of the physiological effects of weapons. And some gases, mainly tear gases, allowed in his view for war to be made more humane and less deadly. More than three decades later, and after the first use of atomic weapons against civilian populations in Hiroshima and Nagasaki, U.S. Brigadier General J. H. Rothschild also pointed out the moral and tactical military advantages of using non-lethal epidemics to take control, with barely any harm to a population, of their territory:

> Surely it is absurd to insist that the savageries of conventional warfare can be even relatively 'humane'. As for nuclear warfare, it should not by now be necessary to emphasize the human damage it can do – not only to the living but to future generations. Ironically enough, it can be argued that the only hope for relatively human warfare in the future lies in the chemical and biological weapons. As I shall try to show, there are gases and biological agents which make it possible to temporarily incapacitate the enemy, with no aftereffects, or to reduce his food supply without indiscriminate killing and maiming. War will never be less than horrible but chemical and biological warfare offer at least some small hope of carrying it on without unnecessarily destroying large numbers of troops, their families, and their cities.[44]

These arguments do realistically address the issue of the physiological effects of a reasoned use of weapons selectively targeting physiological functions. After all, the increasing importance given to avoiding collateral damage in war and the proportionality of the use force being central to international law, why would weapons technically allowing less deadly military operations be more severely banned than less discriminatory or precise ones, such as conventional weapons, causing more collateral damages and relatively unpredictable civilian casualties? Although I do not intend to justify or advocate in this chapter the use of any weapon, it is worth noting that while selecting targets to be destroyed through exploiting physiological vulnerabilities may be considered barbaric or unsoldierly, it is also very much in tune with an explicit goal of modern warfare, frequently put forward by the media or the military (with the metaphor of 'surgical strikes'): to diminish casualties and destruction by better selecting targets.

In international law, the common criteria used to legally distinguish chemical and biological weapons from conventional weapons provide further reasons to unify them under an umbrella concept. The international bioethical criteria in use to morally evaluate weapons, on which policies are grounded, are the vague ones of *maux superflus* (superfluous injury) and unnecessary suffering,[45] which are both medically and militarily defined. In his book *Bioethics and Armed Conflict*,[46] ethicist Michael Gross argued in this sense that these criteria are insufficiently grounded and do not make a significant quantitative or qualitative difference between the effects of weapons affecting physiological functions and those of conventional weapons:

> To identify unnecessary suffering and superfluous injury with what 'most people' accept assumes that most people, on careful reflection, will tolerate the harm that conventional war causes. . . . CBW, by many accounts, will kill far less than one in three of those affected. Nor does it necessarily cause a lingering death, permanent disability, or disfigurement. Detection and protection also offer combatants a fighting chance to survive.[47]

He further explains the general reluctance to admit weapons targeting specific parts of the human anatomy, physiology or biochemistry, as qualitatively similar to conventional ones, by a general cultural aversion to poison and disease, and the fear that physicians may act malevolently:

> Using disease or chemical compounds developed with medical expertise to inflict illness is not qualitatively different from using high explosives to inflict injury and unless they breach the quantitative measures that define unnecessary suffering, neither violates the laws of humanity. Therefore, scientists and physicians should be free to develop both, unless the CB weapons pose the threat of mass destruction.

We shall not here further discuss the idea that the effects of physiological weapons are at least qualitatively and quantitatively comparable to conventional weapons with regard to the harm they may cause on individuals in targeted populations, but further conclude that what biological and chemical weapons, and potentially all agents, inert or alive, able to inflict harm invisibly could fall under a common legal category. This would be so on the grounds that harm is produced in their cases from the basis of medical expertise: the functioning of physiological systems has to be understood in order to select such agents as weapons. While medical expertise is mobilized after the use of conventional weapons to treat war wounds, it is instead required to intervene before the use of physiological weapons, and required even to assess whether the substances or microorganisms used are relevant as weapons. In this sense, physicians and physiological functions play a core role in defining physiological warfare agents.

CONCLUSIONS: FROM PHYSIOLOGY TO THE ENVIRONMENT

To what extent could and should this bodily understanding of warfare be extended to the environmental scale? As stated earlier, biological and chemical warfare do not target individuals but collections of individuals, like any war is in fact comparable to a collection of murders. Thus, in acts of biological or physiological warfare, living organisms are specifically targeted not only individually but also collectively, through the help of medical expertise. If the environment could be used and considered as a potential vector to intentionally produce diseases, should not a comprehensive legislation on physiological warfare include it as such? Does the notion of ecological warfare, put forward by Clarke in *We All Fall Down*, make sense? It seems at this stage hard to exclude the possibility that a physiological warfare could take the form of a poisoning of environments used by humans. This possibility was in fact envisioned in the French CBW programme, in the form of an infected earth policy.[48] One of the unintended consequences of British trials on Gruinard Island, for instance, was the enduring contamination of the grounds with anthrax.[49] This issue has also been regularly raised by military scientists involved in biological and chemical weapons programmes, as well as by pacifists or environmental activists. It is also worth noting here that meteorology has been, since the first endeavours in the area or biological warfare, a key element in biological warfare programmes, as the course of an epidemic is highly dependent on environmental and atmospheric conditions.[50] Moreover, research programmes on herbicides, insecticides or pesticides have been closely linked throughout the twentieth century with military research on chemical and biological weapons.

The link between some agricultural practices and the unintended destruction of elements of ecosystems raised awareness in the 1960s, notably with Rachel Carson's revelation of destructive side effects on birds arising from the use of the dichlorodiphenyltrichloroethane (DDT; an insecticide) in agriculture,[51] in her famous book *Silent Spring*. A little later, and after Clarke's formulation of 'ecological warfare', the notion of *ecocide* emerged, characterized and detailed by Barry Weisberg, in *Ecocide in Indochina: The Ecology of War* (1970).[52] Weisberg also denounced the intentional massive destruction of the Vietnamese jungle with Monsanto's *agent orange* by the U.S. military during the Vietnam War. Botanist and bioethicist Arthur W. Galston, cited by Weisberg, who created defoliants used in Vietnam and developed the *agent orange*, suggested at this time the creation of a new international crime of ecocide: the intentional destruction of natural environments.

Aside from the BTWC and the CWC, a third set of legislations bans such use of the environment for military purposes, among which the first and most emblematic ENMOD (Convention on the Prohibition of Military or Any Other Hostile Use of Environmental Modification Techniques, 1977)[53] bans 'any technique for changing – through the deliberate manipulation of natural processes – the dynamics, composition or structure of the Earth, including its biota, lithosphere, hydrosphere and atmosphere, or of outer space'.[54] But this type of military action and use of chemical agents such as defoliants, designed to selectively target elements of biotopes, fall neither within the category of biological warfare, nor within the one of chemical warfare, as currently defined (as the target of chemical warfare is restrained to human physiology). Despite the difference pointed out by Smith (see the 'Current Definitions' section) between contagions, which are produced through the replication in the host of a biological agent, and intoxications, which require a full dose of a chemical agent per individual to occur, the use of toxic chemicals in an environment can also produce secondary cases, unintentionally, as was the case of dioxin in Vietnam.[55] This contamination need not be intentional to produce an epidemic through the food chain, or other means, which could produce collective intoxications.

On the whole, compartmentalizing aspects of selective destruction of the living for military purposes is not, in my view, an efficient and relevant way of addressing this issue and of protecting human populations and natural environments, as it does not make sense of the common purpose of targeting selectively physiological functions.

NOTES

1 Jeanne Guillemin, *Biological Weapons: From the Invention of State-Sponsored Programs to Contemporary Bioterrorism* (New York: Columbia University Press, 2005).

2 National Academies Archives, Series 1: W.B.C. ('War Bureau of Consultants') Committee, 'Merck Report to Secretary of War: "Biological Warfare"', 3 January 1945, available at http://www.nasonline.org/about-nas/history/archives/collections/organized-collections/1945merckreport.pdf (accessed 27 April 2018).

3 The French case is mobilized here because France developed one of the earliest biological and chemical warfare programmes (before World War I), and the author specialized on the history of this specific programme.

4 The original French reads: 'Cette définition qui permet d'englober dans les armes bactériennes des produits biologiques non vivants, tels que des agents toxiques, rapproche par là même la guerre biologique de la guerre chimique. Il est incontestable que l'utilisation de toxine botulique, par exemple, pourra déterminer des cas plus ou moins nombreux de botulisme mais on ne pourra pas assister à l'évolution d'une véritable épidémie, c'est-à-dire l'apparition de cas secondaires nés des premières atteintes, résultats que l'on pourrait être en droit d'attendre si l'on avait utilise des germes pathogènes vivants' [French Ministry of Defence archives] CAA, Médecin-Colonel Jude, 'La guerre biologique', Archives de la STAT, June 1950.

5 G. Bouthoul. *Le Phénomène-Guerre* (Paris: Payot, Petite bibliothèque Payot, 1962).

6 Ibid. (my translation).

7 Guillemin, *Biological Weapons*.

8 Etienne Aucouturier, *La Guerre Biologique. Aventures françaises* (Paris: Editions Matériologiques, 2017).

9 Etienne Aucouturier, 'Auguste Trillat: Épidémiologie et immunologie de guerre', *Revue d'histoire des sciences*, 67, no.1 (2014), 115–150.

10 Aucouturier, *La Guerre Biologique*.

11 U.S. Department of Defense. *Dictionary of Military and Associated Terms*, 8 November 2010, as amended through 15 October 2015, available at http://www.jcs.mil/Doctrine/.

12 Ibid.

13 Ibid.

14 Ibid.

15 Ibid.

16 Ibid.

17 Ibid.

18 Robin Clarke, *We All Fall Down: The Prospect of Biological and Chemical Warfare* (London: Allen Lane, 1968).

19 See, for instance: Barry Weisberg, *Ecocide in Indochina : The Ecology of War* (First printing. San Francisco: Canfield Press, 1970).

20 C. E. Gordon Smith, 'In Defence of Defensive Warfare', *Nature*, 219, no. 5153 (August 1968), 537–539.

21 Aucouturier, *La Guerre Biologique*.

22 See https://www.opcw.org.

23 Seumas Miller and Michael Selgelid. 'Ethical and Philosophical Considerationof the Dual-Use Dilemma in the Biological Sciences', *Science & Engineering Ethics*, 13, no. 4 (December 2007), 523–580.

24 Ibid.

25 J. P. Zander and K. Nixdorff, 'Enforcing Non-proliferation: The European Union and the 2006 BTWC Review Conference', *Chaillot Paper*, no. 93, ISS (November 2006).

26 Ibid.
27 United Nations Security Council, 'Resolution 1540 (2004)', adopted by the Security Council at its 4956th meeting on 28 April 2004.
28 Jeanne Guillemin, *American Anthrax: Fear, Crime, and the Investigation of the Nation's Deadliest Bioterror Attack* (New York: Macmillan, 2011).
29 United Nations Security Council, 'Resolution 1540 (2004)'.
30 Jeffrey Lockwood, *Six-Legged Soldiers : Using Insects as Weapons of War* (Oxford, NY: Oxford University Press, 2009).
31 John M. Kistler, *Animals in the Military: From Hannibal's Elephants to the Dolphins of the U.S. Navy* (New York: ABC-CLIO, 2011).
32 World Health Organization, 'Health Aspects of Chemical and Biological Weapons : Report of a WHO Group of Consultants' (Geneva: World Health Organization, 1970); Jonathan D. Moreno, *Mind Wars: Brain Research and National Defense* (New York: Dana Press, 2006).
33 John R. Searle, 'Intentionality and Its Place in Nature', *Synthese*, 61, no. 1 (1 October 1984), 3–16.
34 See the Monterrey WMD terrorism database (http://wmddb.miis.edu); S. Wessely, K. C. Hyams and R. Bartholomew, 'Psychological Implications of Chemical and Biological Weapons', *British Medical Journal*, 323, no. 7318 (20 October 2001), 878–879.
35 P. Sartre, 'La légitimité des petites armes', *Études*, 404, (February 2006), 177–187.
36 Jack London, 'An Unprecedented Invasion', *McClure's Magazine*, July 1910. (In this early twentieth-century futuristic novel, Jack London stages an attack on China by the Western world, involving biological weapons, allowing the outnumbered Western countries to win the battle by surprise: the Chinese are puzzled to see enemy planes dropping little glass tubes from the skies, and before they know it, they are defeated by deadly epidemics.)
37 *Panic in the Streets* (1950), *Banco à Bangkok pour OSS 117* (1964), *The Omega Man* (1971), *Virus* (1980) or the more recent *28 Days Later* (2002), *I Am Legend* (2007) or *Contagion* (2011) also emphasize the contagious aspect of BW, by staging deadly and catastrophic epidemics, usually attributable to an intentionally or mistakenly spread engineered and highly deadly virus. BW (in particular, the Japanese human experiments done in occupied Manchuria during World War II) has also inspired musicians. Iron Maiden's *The Breeding House* (1994) and Slayer's more recent *Unit 731* both refer to a Japanese laboratory established in 1941, intended to improve the Japanese CBW programme by testing pathogens and chemicals on local Chinese populations and Allied prisoners of the war in the Pacific. See P. Williams and D. Wallace, *Unit 731: The Japanese Army's Secret of Secrets* (London: Hodder and Stoughton Limited, 1989).
38 Guillemin, *Biological Weapons*.
39 Robert Harris and Jeremy Paxman, *A Higher Form of Killing : The Secret History of Chemical and Biological Warfare* (New York: Random House Trade Paperbacks, 2002).
40 Service Historique de la Défense [French military archives], Ministe`re de la Guerre, 'Instruction sur l'arme chimique', Dossier 9N287, 1926.

41 J. B. S. Haldane, *Callinicus: A Defence of Chemical Warfare*, 1st ed. (London: Kegan Paul, Trench and Trubner & Co., 1925).
42 Ibid.
43 Ibid.
44 J. H. Rothschild, 'Germs and Gas: The Weapons Nobody Dares Talk About', *Harper's Magazine*, 218, no. 1309 (1959), 29.
45 Geneva Convention (Protocol I, 1977a, Article 35).
46 Michael L. Gross, *Bioethics and Armed Conflict: Moral Dilemmas of Medicine and War*, 1st ed. (Cambridge, MA: MIT Press, 2006).
47 Ibid.
48 Aucouturier, *La Guerre Biologique*.
49 R. J. Manchee, M. G. Broster, I. S. Anderson and R. M. Henstridge, 'Decontamination of Bacillus Anthracis on Gruinard Island', *Nature*, 303, no. 5914 (1983), 239–240.
50 Aucouturier, 'Auguste Trillat'.
51 Rachel Carson, *Silent Spring* [1962], 40th anniversary ed. (Boston: Houghton Mifflin Co., 2002).
52 Weisberg, *Ecocide in Indochina*.
53 'Convention on the Prohibition of Military or Any Other Hostile Use of Environmental Modification Techniques', United Nations, 10 December 1976.
54 Ibid.
55 A. D. Ngo, Richard Taylor, Christine L. Roberts and Tuan V. Nguyen, 'Association between Agent Orange and Birth Defects: Aystematic Review and Meta-Analysis', *International Journal of Epidemiology*, 35, no. 5 (1 October 2006), 1220–1230.

Chapter 5

Opening Spaces through Exhibiting Absences: Representing Secretive Pasts

Brian Rappert, Kathryn Smith and Chandré Gould

Many chapters of *Chemical Bodies* illustrate how the history of the use of chemical agents as methods of control and coercion has been intimately tied to the rendering of harm as (in)visible. While suffering has been foregrounded to make the case for brutality and exceptionality on some occasions, on other occasions, suffering has been downplayed, denied or backgrounded to make the case for benevolence and normality. This suggests the need for caution about what is and is not included in any accounts of chemical agents.

Another source for caution is the way the development and use of chemical agents is often undertaken in conditions of secrecy. As a result, scholars, journalists, activists and others investigating such capabilities often take their task as one of exposing hidden truths or unappreciated events. The promise of revealing or unmasking offers a fetching allure for investigators and audiences alike: an invitation to become complicit in a shared but still exclusive understanding.

Against the aforementioned points, this chapter takes the chemical and biological warfare (CBW) programme established under Apartheid South Africa (code named 'Project Coast') as a topic for attending to our commitments in representing the past. This secret military programme used an elaborate array of front companies to camouflage its activities to those both outside and inside of it. Through the endeavours of the Truth and Reconciliation Commission (TRC) and other investigations, Project Coast has come to symbolize the perversities of Apartheid. And yet, each attempt to determine what took place has been delimited by the very terms of the investigations setup to establish the truth. Thus, any attempt to present this programme needs to find ways of acknowledging the partiality of what accounts can be fashioned.[1]

As part of the transition away from Apartheid, over the last two decades, South Africa has figured as a prominent, if not exemplar, case of how regard for the past might encourage reconciliation. By the time of the writing of this chapter, however, the fraught experiences with transitional justice in this nation suggest that pinning down more details about yesterday or providing social meta-narratives through a time-bound truth commission are not sufficient to ensure a just tomorrow. Nor may they satisfy a new generation.

Minding such considerations, our chapter asks: What kind of histories should be told of Project Coast today?

We consider this question in relation to two material forms of storytelling undertaken by the authors: a monograph titled *Dis-eases of Secrecy: Tracing History, Memory and Justice* and an exhibition titled *Poisoned Pasts: Legacies of South Africa's Chemical and Biowarfare Program*.[2]

In this chapter, we consider how these stories of the past sought engagement with what remains outside of them. We seek to move beyond simply orientating to the restrictions on what can be told as information barriers that lead to 'knowledge gaps' resulting in more or less flawed histories. Taking our inspiration from (i) work in geography which understands space not as some fixed terrain but instead as an entanglement constructed from acts of connection and separation,[3] (ii) the 'spectral turn' in the humanities and (iii) 'difficult heritage' practices more specifically, we ask how notions of presence and absence can be intertwined to offer novel possibilities for representing secretive pasts. Our engagement with the relation between presence and absences in these mediums is offered to promote consideration of the purposes of histories, the techniques of rendering bodies (in)visible, the demands of investigating the past and the commitments of unearthing.

CONCEALMENT AND DISCLOSURE

Since its official closure in 1995, Project Coast has been the subject of a number of inquiries and publications.[4] Notably the TRC examined it through public hearing in the late 1990s. The TRC hearing was the first large-scale public exposure of a national chemical and biological weapons programme.[5] Also, between October 1999 and April 2002, Dr Wouter Basson faced sixty-seven fraud, murder and drug charges directly or indirectly related to his activities as the long-time head of Project Coast. This was one of the longest trials in South Africa's history. Basson was not found guilty in relation to any of the charges. This trial was the subject of a book titled *Secrets and Lies* co-written by Chandré Gould. In addition, in 2002, Gould co-authored a 300-page report for the United Nations Institute for Disarmament Research and Centre for Conflict Resolution titled *Project Coast: Apartheid's Chemical and Biological Warfare Program*.[6] In

2013, the Health Professionals Council of South Africa (HPCSA) concluded a fraught and lengthy disciplinary hearing and found Basson guilty of unethical conduct for activities while he was head of Project Coast. As of late 2017, a sentence had yet to be passed. Basson is the only person associated with the programme to have been legally charged or professionally sanctioned. Through these investigations and others, much has been publicly disclosed about Project Coast. And yet, with this disclosure, it has also become plain that much is not understood, and may never be so.

In this section, we indicate some of the ways in which concealment and disclosure have characterized *both* the activities in the programme and the investigations into it to determine what took place.

To begin, concealment was a sought capability within the activities of Project Coast; particularly the development of undetectable ways of chemically killing and injuring. Among the weapons that did not look like weapons included spring-loaded screwdrivers, bicycle pumps and umbrellas that released chemicals of choice. Finger rings with hidden poison compartments were manufactured. In combining the right-looking 'applicators' with the desired acting substances, many variations were possible: chocolates with botulinum toxin, anthrax-contaminated envelope glue and so on.

When understood as assassination weapons, such applicators disguised purpose in appearance. Upon accusatorial questioning and TRC hearings in 1998, however, individuals such as Wouter Basson repeatedly defended their production, suggesting the devices' real purpose was not what it seemed. What conceivable reason could there be in putting cyanide in peppermint chocolate then if not to poison? To demonstrate to members of South African security forces what Russian trained African National Congress (ANC) operatives could do with simple chemical substances, Basson retorted. Why lace cigarettes with anthrax? To test if this delivery system could harm people (tests showed it could not). Why produce bottles with cholera? To test inoculation processes and advise about how to deal with potential outbreaks in Namibia.[7]

Investigations into this programme were also characterized by a dynamic interrelation between disclosure and concealment. Particularly in the initial years after Apartheid, technical, scientific and operational documentation associated with Project Coast was scarce. Records of this closely guarded and officially secret programme were not available to the public or even some official investigators during Apartheid and in the immediate years thereafter. Even those who may have had a right to know – such as TRC investigators – were effectively barred from military archives, not least through the requirement that they had to make specific requests for particular documents despite not having any knowledge of what records were kept or how the South African Defence Force (SADF) archival system operated.

Instead of deriving from a general opening up of the security state, the large majority of the documents that proved central to later investigations came indirectly and serendipitously. To detail one chief set of eventualities, on 29 January 1997, South African Narcotics Bureau detectives set up a sting operation in a parking lot near the home of Dr Basson. He turned over a black plastic bag with 1,040 capsules of the street drug Ecstasy to a business acquaintance who then placed 60,000 rand on Basson's car seat. The operation eventually led the police to the home of an associate of Basson. There, two blue sealed steel trunks were located and subsequently two others were discovered. They contained more drugs, as well as hundreds of sensitive and classified documents that would become central to constructing accounts of what took place.[8]

Furthermore, while eleven witnesses associated with the programme testified in the TRC public hearing, three senior figures did so reluctantly. TRC panel members sought to question witnesses about complicated financial, managerial and technical matters but with only the limited documentary evidence in their possession. For their part, witnesses often struggled with remembering events of years past but also making sense of documents and activities they reported not knowing about.

Likewise, the criminal case against Basson was plagued by difficulties about the limitations of evidence.[9] For many murders or chemical interrogations in which Basson was alleged to have played an active role, the only witnesses present were those that had carried them out. As such, their testimony before the court was open to question about their motivation. In general, these witnesses were regarded as flawed by the trial judge. Many of those associated with the SADF could not have been called by the prosecution because they would have been hostile, more likely to harm than help the state's case. Moreover, Basson's trial was not set up to hear – let alone resolve – disputes about activities raised in previous investigations but that were not connected to the specific fraud, murder or drug charges. As a result, activities such as the authorization and development of anti-fertility vaccine to be used against black women discussed at the TRC were not examined at the trial, nor in the HPCSA professional hearing. Indeed, the HPCSA pursued charges based only on what Basson himself had admitted and testified to in the trial, such as the production of drugs and tear gas, in an attempt to prevent lengthy legal contestation.

Reluctance to air evidence openly was not just evident by those individuals and organizations cast as wrongdoers. At the TRC, and subsequently, officials in the South African ANC-led government would warn about the dangers of publicly detailing Project Coast. Why would the post-Apartheid government not want to publicly reveal details of a chemical and biological weapons programme that targeted its own members? Besides the proliferation dangers of airing information, it is likely that the newly elected ANC government would have feared that something 'unknown', and potentially embarrassing or harmful, would be revealed. Particularly in the early years after Apartheid, the information the new state had about the programme was only that which they had been given by Basson and others associated with Project Coast. The extent to which that picture was complete could only be guessed at by ANC government officials. In other words, what 'the state' itself was 'allowed' to know was limited to what it was officially told. As such, the fledging ANC government needed to protect the secret that it did *not* know.

The investments made into believing that highly dangerous secrets existed and that they needed extraordinary protection measures was evident in relation to the notional method for retaining technical documentation. As part of the winding down phase, the management committee for Project Coast decided certain documentation should be saved electronically on CD-ROM discs while the physical versions would be destroyed. Especially given the sealed steel trunks found in 1997 at the home of an associate of Basson, what had happened to the electronic versions of Project Coast documentation took on some importance subsequently. Surgeon General Niel Knobel (head of Coast's Management Committee) spoke in detail about the elaborate provisions made to keep this information secure at the 1998 hearings:

> After the technical information was transferred from documents onto the discs, the discs were brought to me by Colonel Ben Steyn [the last Project Officer of Project Coast] in a safe. I established that the discs were inside the safe, and as far as I remember there's also an additional floppy along with it, which is the access mechanism, access coding that you require to be able to access the information on the discs.
>
> It was then put into a very large wall safe attached to my office and my headquarters and only Colonel Steyn and I had control, joint control over the small safe, smaller safe, the portable safe. After the demarche and particularly after the Americans and the British expressed concern about the safety of the information on the discs, I went to see [South African President] de Klerk and I followed it up with a letter and that letter I can give you a copy of. It was in April 1994.
>
> At that stage we changed the joint control in such a way that all three of us, the President, Mr de Klerk, Colonel Steyn and myself had to be present in order to access or to be able to open the small safe. The position was then changed, it was then changed to a safe in a different part of my headquarters, a huge safe with two keys and a combination and the small safe with its two keys was put into the bigger safe.
>
> And in that joint control we gave the President one of the keys of the big safe as well as the combination of the big safe. I kept the key of the big safe and one of the keys of the small safe. Colonel Steyn had the combination of the big safe and the other key of the small safe, and that was how that situation was maintained.[10]

As subsequently recounted by Surgeon General Knobel, when the new government came to power in 1994, the key and combination given to President de Klerk was passed over to Deputy President Thabo Mbeki. Yet what was actually on the discs, and why it was retained was not clear, even to those who held the keys.

In openly discussing the existence and whereabouts of the discs, the media and the TRC posited that there was something on the discs that needed to be hidden because of its profound importance.[11] As Burger and Gould noted in *Secrets and Lies*, the 2002 book about Basson's criminal trial, a later review undertaken of the 'CD-ROMs stored as a "national asset" under the most stringent security concluded that, far from being the repository of Coast's deepest, darkest secrets, the discs contained little more than published literature on [chemical and biological warfare] in general'.[12] While the existence of the discs was discussed during the TRC hearing, and in *Secrets and Lies*, the power of the push to retain secrets was made evident in a 2013 interview between two of the authors (Gould and Rappert) and a senior member of the military. He did not wish even to mention the existence of the discs on record because he felt they were still too sensitive.[13]

In a similar vein, the light cast by investigations might well be judged as not always illuminating. Because the Basson trial was an undertaking to consider criminal charges against him as the only accused, possible wrongdoing by others in Project Coast was sidelined. As head of the programme for most of the secret military programme's existence, this is unsurprising at one level. However, the focus on the individual, and thus the personification of the programme both during and subsequent to the trial, arguably has served to distract attention away from the many others who held leadership positions in the military, or even in the companies that made up Project Coast, and those who performed the more mundane tasks that allowed the programme to function. To this day, the focus on Basson alone allows, even encourages, a perception that the criminal trial and the HPCSA's hearing were 'witch hunts' that unfairly targeted the most visible, most exposed person.[14]

A SENSE FOR ABSENCE

The previous section gave a flavour of the ways in which what has and can be said about Project Coast implicates a sense of what remains unsaid. Such considerations undercut the belief that some definite and definitive account can be given. We propose that one sensibility that follows from the analysis earlier is that rather than being sharply divided, disclosure and concealment should be understood as subtly interrelated.

The need to recognize this is evident in how histories to date have been positioned. For instance, in 2002, some commenters drew the lesson in the report *Project Coast: Apartheid's Chemical and Biological Warfare Program* that 'having come clean on its experience during the Apartheid years lends real credibility to South Africa's ethical and practical stance on international disarmament. South Africa went to the edge and beyond and then – under a new, enlightened regime – came back. Others can do the same'.[15] Whether such optimistic appraisals can be justified today is open to doubt.

Or, at least, that is the way the authors of this chapter would regard the situation. The sensitivities surrounding what histories can and have been told, in turn, signal the importance of minding wider cautions about what can be expected from history. Having undertaken significant efforts to detail the Apartheid past, we would regard the aspiration to 'set the record straight' as problematic. These are some of the reasons why:

(1) Experience to date with Project Coast has indicated the ways in which disclosures provide the very basis for yet further questioning. One example would be the revelation by a former SADF soldier, Johan Theron, during the Basson trial that hundreds of sedated South West African People's Organization (SWAPO) prisoners of war had been thrown from an aircraft into the sea off the Skeleton Coast. To date, all attempts to verify this information, or seek further information that could reveal the identity of those killed, in this way have failed, including attempts to engage the assistance of the Namibian authorities. This raises questions about the veracity of Theron's claims, and about the reluctance of the Namibians to resolve the mystery.[16]

(2) Despite the often held belief that revelation leads to lessons that can work to prevent future abuses, this is not necessarily the case. A 2015 study by the Academy of Sciences of South Africa found that 'education and/or training on research ethics, including issues such a scientific misconduct . . . is not routine for life scientists' and 'there was a low level of awareness among life scientists about national and international conventions, laws and regulations related to their research; and that information about these instruments is not readily available'.[17] This was so despite many years of work, after the TRC hearing, to raise awareness about chemical and biological weapon programmes among scientists, academics and professional associations and draw lessons about the importance of ensuring that the training of scientists includes reference to the international norms (as enshrined in the Biological Weapons Convention (BWC) and the Chemical Weapons Convention (CWC)) as well as national legislation.

(3) In South Africa, in particular, the bargain of truth in exchange for amnesty from prosecution within the TRC did not work quite as well as hoped. As discussed in Lentzos's *Biological Threats in the 21st Century*, 'there was little or no motivation for amnesty to be sought in cases where the perpetrators were fairly certain their deeds would remain hidden, such as information about the details of military involvement in human rights violations. Indeed, the military closed ranks, effectively boycotting the TRC process, with only a handful of amnesty applications being received from soldiers'.[18] This suggests a need to rethink investments placed in 'truth'.

(4) Also, as became clear in recent dialogues hosted by the Nelson Mandela Foundation with memory workers from around the world, formal transitional justice processes and efforts at 'truth telling' do not necessarily lead to healing as may have been presumed.[19] Indeed, the way in which such processes reinforce a binary between 'victim' and 'perpetrator' may well create the conditions for the reoccurrence of violations by 'righteous victims'.[20]

(5) Processes of revelation based on investigation necessarily excluded much of the information that is gathered. Not only is it simply not possible, or in some instances unethical, to include all information gathered during an investigation; it also is necessary to create narratives that are understandable and useful to intended audiences. As such, investigators have to exercise discretion and caution in deciding what is put into the public domain, at least if the intention is to minimize further harm.

(6) Finally, one of the difficulties of investigating Project Coast relates to the overabundance and familiarity of stories and evidence. Such a situation of surfeit presents its own potential to render some matters obscure. The question then for representing Project Coast now is 'How can we effectively open space for something new?'.

As a result of these considerations, what remains unknown is not simply this or that remaining piece of the puzzle, the placement of which can complete the picture of the past if only we could unearth it. Instead, varying suggestions about what still needs to be established and who possesses such information helps constitute radically alternative understandings of the outline, morals and lessons to be derived from the past.

Rather than seeking to pry open the secrets hidden away, our experiences to date with the production and reception of accounts of Project Coast suggest adopting a sense of alertness to the expectations from history. To put it in other terms, what is called for is more aligned with inquiry rather than resolution, receptiveness to possibilities rather than detailing particulars. In the remainder of this chapter, we elaborate ways sought to realize such an orientation. We consider possibilities for negotiating presence-absence in relation to two different types of interventions: a co-authored book *Dis-eases of Secrecy* by Rappert and Gould as well as an exhibition curated by all the authors titled *Poisoned Pasts*. These were not efforts to 'set the record straight' about Project Coast, but instead efforts to attend to the manner in which histories are made and remade by using connections and separations to create alternative spaces.

DIS-EASES OF SECRECY: SEWN THREADS

A basic intent of the 2017 book *Dis-eases of Secrecy: Tracing History, Memory and Justice* was to cultivate an awareness of the conditions for knowing; this by writing in a format whereby we as authors sought to develop a sense of the fractured and fraught process of assembling a history of secreted events – secreted in the both senses of the term: released and concealed.[21]

This was done through an explicit two-part organization. The entire main text is broken up into 548 numbered entries that are given in roughly chronological order. Each entry provides its own fragment account of the past. The entries can be read from front to back in the conventional fashion, though with some effort. In addition, the entries are organized into eleven themes. Entries one to eleven in *Dis-eases of Secrecy* provide the first entry for each of the eleven themes. At the end of these starting entries, readers then are given the number of another entry that continues that theme. They carry on reading until directed to another numbered entry, and so on moving back and forth in the book between different time periods. As the reader progresses through the text, they find themselves circling back on entries they have encountered before but situated in a different sequential context. The circularity of the thematic organization itself points to the lack of resolution or a clear end. The intent was to mirror the process of inquiry and investigation into troubled pasts. In other words, the flipping back and forth between pages to read the entries is a physical embodiment of how the assembling of histories takes place, with connections being made between some events and not others.

The intellectual inspiration for this organization was Sven Lindqvist's *A History of Bombing*. This entry-to-entry technique enabled Lindqvist to produce an absorbing, detailed, yet accessible overview of the fantasies, expectations, terrors, fascinations and duplicities of aerial bombing. Through the thread-based entry into the 'labyrinth' of past, he encouraged readers to sense that they are only taking 'one of many possible paths through the chaos of history'.[22]

We too were interested in opening options that enable a tracing of the past. Our orientation in relation to the entangled past of Project Coast differed somewhat from *A History of Bombing* in that we were often not sure about how to characterize what took place – whether that be a labyrinth, scheme, network, circus or cycle. Our themes were not single-track lines for getting through a complex labyrinth. Instead, by offering discourses that intersect, branch off, get read in reverse order and so on, we sought to stimulate imagination for ways of diagnosing situations anew. We also sought to place the reader amid the sometimes contradictory and confusing interpretations from multiple sources to enable them to experience the choices associated with writing about contested events. For instance, an entry positioned as part of the sequence of one theme can take on a rather different meaning when positioned elsewhere.

The goal, in short, was to turn a technique for reading into a method for inquiry. We wrote with absences and secrecy to alert readers to the processes of producing knowledge of the past. Through an argument in which what was missing was intended to be as much a feature of what was given, we sought to encourage readers to partake in a process of investigation – with the lures, dissatisfactions and affects that this can entail. It is between the traces and fragmentations of our accounts that we hoped readers would be able to ask questions of themselves and others about the purposes of history and the potential of memory. In doing so, we hoped to engender sensitivities with the investments in the telling of history.[23]

POISONED PASTS: MATERIAL LACUNAE

In 2015, the authors partnered with the Nelson Mandela Foundation in the development of an exhibition titled *Poisoned Pasts* that took as its central theme the Apartheid-era chemical and biological warfare programme. It initially ran at the Nelson Mandela Foundation Centre of Memory between October 2016 and July 2017 and at the time of writing was housed at the Steve Biko Centre (Ginsberg Township, Eastern Cape).

Poisoned Pasts sought to stimulate questions and discussion about the responsibilities of scientists to prevent the malign application of science, to consider how easily scientists might be drawn to similar work in the future and to question the closure promised by justice and transitional justice related to South Africa. The exhibition was offered as a response to the frustrations of transitional justice to deliver accountability and sanction of those involved in human rights abuses, with a particular goal to engage (younger) audiences in questions about the relevance of the past to the present.

As in the case of *Dis-eases of Secrecy*, as the curators of *Poisoned Pasts*, we sought to do this in part by using the interplay between what was displayed and what was not (or could not be). As a physical space, an exhibition can make use of a far greater range of modalities of expression than a written manuscript. Extending the idea of turning a technique for reading into a method for inquiry, we sought to design an exhibition that would encourage visitors to assume an active investigative role. While this intention was conveyed in the written exhibition guide (a text that may or may not have been read by exhibition goers), it was achieved through the design choices made.

In the remainder of this chapter, we attend to how we attempted to realize our intentions and motivations, first in relation to the question of materiality, covering both artefacts and the construction of exhibition elements, and then in relation to how the spatial design of *Poisoned Pasts* extended these decisions, allowing the play of presence and absence to find both explicit and implicit expression. This leads finally to a consideration of the representation of those affected by Project Coast, and how these visual solutions reflect our desire to acknowledge both those potentially lost to the project and the challenges of accounting for unresolved deaths and disappearances.

Material Objects: Three Registers

The primary (but by no means exclusive) mode of engagement within exhibitions is an aesthetic one. The raison d'être of exhibitions presupposes visual content, including objects that demonstrate and authenticate the events being described. In this respect, designing *Poisoned Pasts* presented a particular challenge. Whether due to wilful destruction, standard procedures, or shoddy bureaucratic management, very few material artefacts survive that can be taken as 'proof' of Project Coast's existence or activities. Only one 'true' artefact of the programme could be located and made available for inclusion in the exhibition: a collection of modified screwdrivers and other fragments from covert devices that were entered into evidence at Basson's criminal trial. What also remains of the project are several thousand pages of archival documents, mainly facsimiles. With this documentation, we are still left with the issue of authenticity, but the more pressing concern is how to render these many pages of (partial, redacted, questionable) data accessible, useful and productive, especially as our primary intention in designing the exhibition was to position visitors as active investigators that could open up new possibilities for received histories.

One way we dealt with this poverty of objects was to fashion proxy objects. These included a 1:1 scale reproduction of a personal safe in which CD-ROMs containing the records of Coast were allegedly stored by Surgeon General Niel Knobel (see above), as well as a custom-designed restraint chair for animal experimentation (probably baboons), used by the project at Roodeplaat Research Laboratories. The chair was reconstructed after an original design by biomedical engineer Jan Lourens; its form and functions were detailed during Lourens's TRC testimony and later interviews with Gould and the exhibition design team. Constructed in thick Perspex, the original unit (current location unknown) consisted of a chair with straps to restrain an experimental animal in a transparent housing that would allow scientists to visually observe the effects of various gases on the animals. For *Poisoned Pasts*, both the safe and the chair were rendered in corrugated cardboard – functionless and impotent. Other proxies included an array of everyday groceries and toiletries that Project Coast scientists sought to modify with various poisons.

We regarded the deployment of proxies as enabling but also tension-ridden. Displaying objects of violence may well invite criticism that the exhibition makers are, wittingly or unwittingly, participating in the extension of trauma through choreographing violence-as-spectacle. Artefacts of real or intended crimes gain an undeniable 'thing power' (to use Jane Bennett's phrase[24]) that in turn invests them with a potency – their lure is hard to resist (the record-breaking attendance at the Museum of London's 2016 exhibition 'Crime Museum Uncovered' bears this out). Too, we cannot avoid another kind of fetishization that comes with an indexical object, the bearer of a forensic 'I was there' performance of authenticity. The benefits of exhibiting such objects should be tested against following questions : What might be gained from doing so? Could exhibiting the objects be harmful? To whom?

Besides the aforementioned strategies, the fictions and absences of Project Coast were communicated by engaging with different modalities of visual representation, from the photographic to the illustrative. For example, *Poisoned Pasts*'s visual language was graphically bold, making extensive use of South Africa's rich legacy of photojournalism. Acid yellow, deep red and ultraviolet blues dominate the colour palette, along with visual filters such as the half-tone screen (the dot-matrix of traditional reproduction technology) and negative inversion created textures from photographic images. These were used as visual cues as they have become inscribed in Western visual culture as semiotic shorthand for mass-media appropriation while also signifying a deconstructionist position in relation to it. Such deliberate (re)mediations make implicit and explicit statements about the limits and constructions of archival practices, and in turn, how the exhibition confronted the challenge of truth-telling and truth-construction through 'real' and simulated objects.

In sum, the objects (including images and documents) in *Poisoned Pasts* therefore operated across three registers: (a) *indexical*: those which bear a direct relationship or connection to the programme or related processes of social/legal justice or professional standards; (b) *indicative:* reconstructed or 'proxy' objects that stand in for objects either lost, destroyed or only available via documented description; and (c) *imaginary:* content that attempts to give visual form to events that cannot be proven or verified but which figure strongly in the public narratives surrounding Coast. These were designed as necessarily ambiguous coordinates that might allow visitors to recall, contest, connect or discover additional perspectives on the programme.

Interrupted Sightlines

Exhibitions are spatialized experiences as much as they present collections of objects, media and interpretive materials. They are spaces we move through bodily, using multiple senses to receive and process what they offer to us, which in turn cannot be completely divorced from what we bring into exhibitions – our personal stories. Successful exhibitions generally present their content – objects, audiovisual media, interpretive material, in other words everything that is explicitly visually available to us as visitors – in close integration with the space in which the exhibition will be experienced. In much the same way that typography can be used to reflect tone of speech (e.g. capitalized words in emails or short message services [SMSs] inferring heightened emotion), an exhibition's spatial design is a potent carrier of implicit communication about how its makers intend its content to be interpreted.

Whether an exhibition of contemporary art or social history, both exhibition makers and visitors agree that the 'book-on-a-wall' approach makes for a pretty fatal exhibition experience, particularly in an age of heavily abbreviated digital communication where our attention spans are largely being trained out of durational, closely focused concentration. Thus, objects available to make an exhibition must be considered in relation to their spatial interpretation. In fact, spatial arrangements such as juxtaposition, vertical hierarchy and order of encounter grant additional affordances to an exhibition's objects. How will they be placed in the exhibition space? What will support them – literally, conceptually and thematically? What, in turn, will they support?

With *Poisoned Pasts*, we actively sought to create spatial conditions whereby details of the Project Coast story may be happened upon, through a non-linear/non-chronological organization of material and multiple pathways through the space.

Figure 5.1 provides the floor layout for *Poisoned Pasts* when it was at Nelson Mandela Foundation Centre of Memory. The design of exhibition was conceptualized in the image-composition terms of 'Foreground' (the bottom area near the entrance), 'Middle-ground' and 'Background' zones, which reflects something about the content positioned within each, which was arranged in terms of three broadly thematic clusters of intersecting panels that should be considered 'in the round'.

The visual inventory of the exhibition consisted of panels of researched commentary with supporting visuals, predominantly visual reportage (photojournalism), historical documents and original artefacts interspersed with proxy objects.

The 'Foreground' included the title/contributors panel alongside a short (five minutes) introductory film *What Was Project Coast*, which offered visitors a core set of facts about the programme while explicitly acknowledging that the narrative relayed was only one of many possible narratives. The rear of this arrangement – 'Declarations and Silences' and 'Hostile Uses of Chemistry and Biology' – asked visitors to consider the current (as of mid-2016) climate of CBW policy and application internationally, and South Africa's declared position: which is that it never had an offensive programme. If visitors decided to look at both sides of the panel first, this statement acted as a position against which to test what was discovered within the exhibition. If visitors proceeded into the space and circled back around and read this last, it may have prompted a reconsideration of the honesty of South Africa's historical position and what this implied for the future.

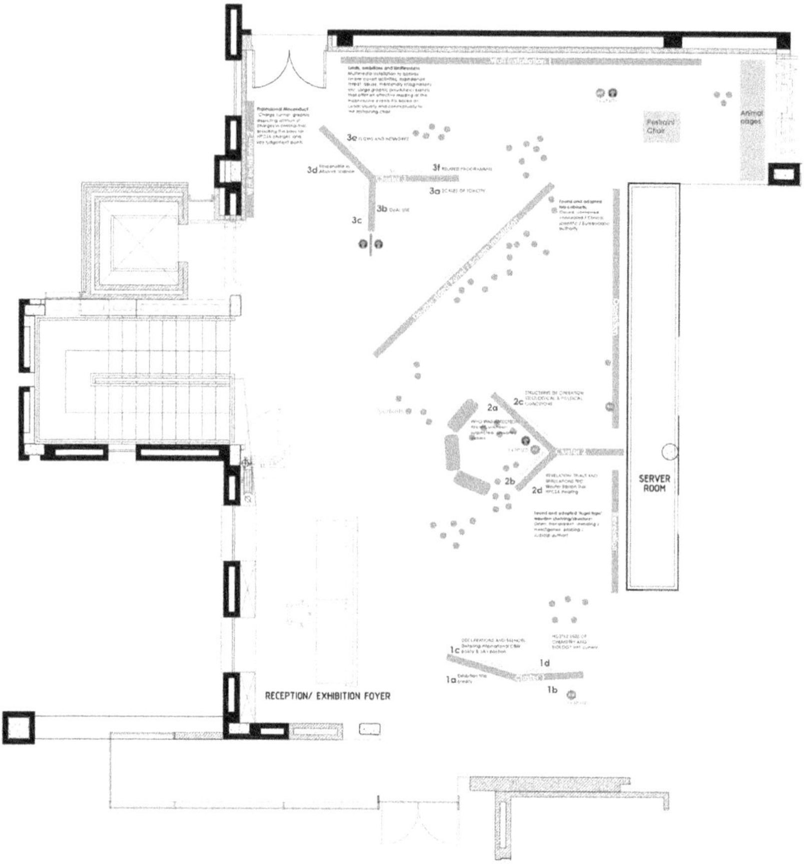

Figure 5.1 ***Poisoned Pasts*** **exhibition design: floorplan for installation at Nelson Mandela Foundation, Johannesburg (Courtesy of TRACE).**

The 'Middle-ground' zone was set against a wall. An alcove created by two intersecting screens created an intimate space in which to consider accounts of 'Who Was Affected?'. A comprehensive narrative list of known, unknown, suspected and 'unworthy' victims was presented, and a small audiovisual station features archive footage of multiple narratives – Sarah da Fonseca telling the story of the suspicious death of her husband, Special Forces soldier Victor da Fonseca; Rev. Frank Chikane recounting his own poisoning; Maria Ntuli, mother of Jeremiah, one of the Nietverdient 10 who were lured away together and killed when the van they were travelling in exploded; and Rev. Peter Kalangula describing various attempts on his life by South African security agents in Namibia.

Facing this alcove, two sets of shelves were visible against the back wall. These demarcated the thematic spaces of 'Operation' and 'Revelation' on the left and right sides, respectively, with supporting illustrated panels covering 'Structures of Operation: Ideological and Political Conditions' and 'Revelations: Trials and Tribulations', including the criminal and professional ethics trials of Dr Wouter Basson.

Allowing for necessary blurring of distinctions between the various zones of foreground/mid-ground/background, 'Operation' and 'Revelation' represented events and documentation in the public record, the public-facing aspects of what was known about Project Coast and chemical and biological warfare within the contemporary international context. These two spaces were physically separated but remain visually connected by a PVC strip-curtain of the kind found in hospitals or slaughterhouses. In addition to its dual connotations of being associated with the preservation and ending of life, the semi-transparent, semi-permeable membrane of the PVC strip-curtain was used because it enabled mediated transparency and movement. Diligent visitors trying to see (and move between) the displays could do so but only with effort. The displays were intended to indicate how concealment and disclosure blended together in the activities of Project Coast and the investigations into them. The texts, objects and audiovisual material for both areas was organized according to headings such as 'Hidden in Plain Sight', 'Locking Away' and 'Silences'.

A large screen on a near-architectural scale divided the rear part of the exhibition in two, on the diagonal, forcing a spatial closing-down of the 'Operation' space and masking what lay beyond. This 4.5-metre panel acted as the boundary between the conceptual 'Middle-ground' and 'Background' zones of the exhibition's spatial design. Entering this area was to travel further back into recent history, where things get slightly less sure-footed in terms of the official record, and perpetrators engage in a selective accounting of their activities. The side facing into the Operations space was boldly visual, depicting the shadow of a light aircraft cast onto a coastline as it flies above, beyond the camera's frame. Striations in the visual reproduction – making an asset of the media artefact known as 'interlacing' – denoted a video source for this image, captioned with testimony given in Basson's criminal trial in which witness Johan Theron (see above) described being involved in the clandestine dumping of sedated bodies of Angolan soldiers into the sea off the Namibian coast.

The reverse of this panel was a graphic visualization of the infamous 'Verkope Lys' ('Sales List') in which descriptions of the poisons on the list are set against various human protagonists and experimental animals, extending the trope of 'agents' through the figures of human, animal and chemical bodies. Here a visitor entered the 'Background' zone, in which Coast is placed within an international and historical context of similar programmes and science's dual uses. 'Background' here has both contextual and covert connotations.

Set against a large-scale relief mural (*Limits and Limitlessness*) designed by Smith and rendered in a palette of (ultra)violets, blues and blacks that references some of the most violent and controversial actions of the programme, the third and final zone was arranged as a set of three intersecting panels and a related stand-alone wall panel to one side, and the reconstructed restraint chair, and a set of cages that once housed experimental animals to the other. The cages were loaned from one of the Project's former lab facilities, which was subsequently purchased by the South African government. Visitors could also listen to 'Science and Society', an audio montage by Smith featuring different perspectives on the roles, responsibilities and culpabilities of scientists associated with the programme, and consider an indicative reconstruction of an abandoned experiment in which scientists attempted to invest maize meal – a staple food – with cyanide. Three bowls of maize meal, tinted varying shades of blue, were presented in yellow enamel bowls of the kind associated with low-income households.[25]

Hung opposite the restraint chair and animal cages, the stand-alone wall panel detailed the outcome of Basson's criminal trial, highlighting key points within the judgement, as well as the attrition of criminal charges that provided the basis for later charges brought by the Health Professions Council of South Africa. Within this zone, an animated film detailed Coast's exchanges and trades with international partners carried out under the auspices of numerous front companies which broke many of the economic sanctions imposed on Apartheid-era South Africa.

One notable feature evident from the layout schematic is the lack of clear sightlines. Within exhibition design, working with sightlines affords curators the opportunity to infer relationships between objects within and across the exhibition environment. Sightlines may be employed as active design elements, or incidental ones. Being alert to them affords the observant visitor an extra, implicit level of interpretation that is open to individual, subjective experience, and which may not reveal itself in the same way twice.

Within *Poisoned Pasts*, however, we worked with sightlines as a way to *disrupt* an easy visual experience. This approach was informed by 'exploded view' drawings:

> An exploded view drawing is a diagram, picture, schematic or technical drawing of an object, that shows the relationship or order of assembly of various parts. It shows the components of an object slightly separated by distance, or suspended in surrounding space in the case of a three-dimensional exploded diagram.[26]

Common to engineering instruction manuals, the exploded-view schema will be familiar to students of anatomy albeit in object-form: 'exploded skull' models show the various bones of the human head separated from each other on rods.[27] This principle of a complex object consists of interlocking parts was an appropriate metaphor for the relationship between content, space and active visitor we wished *Poisoned Pasts* might embody. Yet more than just a pulling apart of objects to reveal their constitutive parts, we sought to reveal both what was given (explicit) and what was not (implicit) to engender a set of possibilities for the exhibition space.

The space of the Nelson Mandela Foundation was configured such that a visitor was unable to take in an overall view of the exhibition at once. As investigations into Project Coast has produced blind alleys, dead-ends and blank walls that occasionally open up into revelation and understanding, only partial views are possible. Thus, in order to see the 'totality' of what is on display, one must assume a willingness to explore, inspect. This placed a particular demand on a visitor; they needed to intentionally manoeuvre through objects, visuals and recordings, instead of following a linear thematic or chronological path. This demand was additionally one of time, a particular commitment to an exhibition experience that exceeds the purely visual. The questioning we wished to prompt was one of 'What's around the corner?'.

Depending on the choices made on the path through the exhibition, some matters would come into view but others will be occulted away. Individual

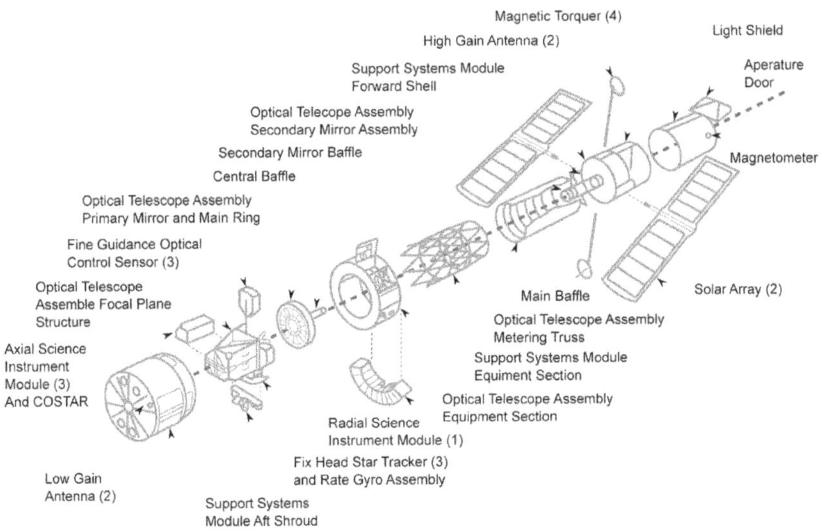

Figure 5.2 'Exploded view' schematic diagram of the Hubble Space telescope by Andrew Buck, 2009. Licensed under the Creative Commons Attribution-Share Alike 3.0 Unported license. https://commons.wikimedia.org/wiki/File:HubbleExploded.svg

journeys through the displays necessarily entailed circling back on material visitors may have already seen, but with an altered background of knowledge. This is an embodiment of how investigations of troubled pasts lack a simple vantage point from which events can be reconstructed.

Form was matched by content because *Poisoned Pasts* did not provide a structured storyline for meaning-making. Other than the introductory video that audiences may or may not have watched, no overview narration was provided. The clusters provide information on particular topics, but without the displays being linked together by any meta-arcs. By this, we encouraged connections to be made across thematic clusters without choreographing connections through forced or overstated juxtaposition.

Material Representations

Narratives that bear witness to previously unrecorded or otherwise-suppressed political violence are a critical feature of transitional justice processes. A preferred tactic for heritage exhibitions is to focus on narratives of those affected by events, perhaps because it mirrors the processes of transitional justice. But what if stories of victimhood are not available, or the nature of such stories cannot be adequately verified, or the boundaries between victim and perpetrator are fluid, subverting our conventional ethics?

Witnessing, as Eamonn Carrabine reminds us, has both a juridical connotation – 'seeing with one's own eyes' – and a religious one – 'testifying to that which cannot be seen . . . bearing witness'.[28] A seemingly endless torrent of powerful victim testimonies enfolded with acts of forgiveness and absolution – and rejections of such conciliatory gestures – characterized the TRC process overall in South Africa. This was so, not least, in the figure of Joyce Mtimkhulu, whose son Siphiwo was poisoned, most likely from thallium. Although this did not kill him, his later disappearance and murder resulted in his body never being found. This lack of material evidence of a life lived was at the core of his mother's demands before the TRC. She appeared before the commission in 1996 with a handful of his hair which she had collected as it fell out, a direct effect of the poisoning. Without bones to bury, she asked, how could she resign herself to his death and do the work of grieving?

In the case of Project Coast, considered responses to the questions 'Who were the victims of the programme?' and 'How did they suffer?' need to be told, at least in part, in the negative. That is, what has not been told, what cannot be told or what will not be told would need to figure prominently. This is so because records were not made, many were destroyed, few witnesses have spoken up, appropriate questions were not posed, a 'need-to-know' principle was deliberately applied to ensure plausible deniability and so on. But the problems at stake go further than a lack of information. Not all those killed or injured as part of Project Coast would likely be regarded as meriting the label of victim for the purpose of memorial histories.

One classification for who was affected by the programme would be as follows:

Identified – There are few identified victims who can be linked directly to the programme. The most prominent individual identified as a victim of poisons produced by Project Coast has been the anti-Apartheid campaigner and former Secretary General of the South Africa Council of Churches, Reverend Frank Chikane.

Nameless – For a variety of reasons, those who were determined to have been assassinated have been largely unidentified. Who would be the target for the concealed weapons like the laced chocolates or envelope glue was something senior scientists could often rarely more than vaguely speculate about because under Project Coast, they were deliberately kept in the dark. For field operatives, at times, targeted individuals less prominent than Frank Chikane were not remembered in name during post-Apartheid investigations. They might be referred to loosely as something like 'friend of black dissident',[29] an 'ANC member' or a 'foreign operative'.

Unpursued – Many of the soldiers claimed to be killed have not had the circumstance of their deaths investigated in detail, and most remain unnamed. A 1989 proclamation (only revealed because of the High Court trial of Wouter Basson) had granted amnesty to all SADFs and others for crimes committed in Namibia prior to its first democratic elections in 1989. Citing this amnesty, the judge in the trial threw out charges against Basson related to the murder of some 200 SWAPO fighters, plans to poison the water supply of a SWAPO refugee camp with cholera and the murder of an official. The other charges were related to conspiracy to murder in Swaziland and Mozambique through the supply of poisons. As these had taken place outside of South Africa, even if they were planned within it, the trial judge ruled they could not be prosecuted in the country.

Numberless – Surviving documentation indicates an extensive range of animals were tested on as part of the programme. To name but some instances, beagle dogs and baboons were administered the pesticide paraoxen in varying mixtures. Rats were given poisons, baboons were given rat poisons. Tests on one species led to confirmation tests being done on another – pigs, rabbits, horses and primates were all studied as models for the effects of chemical on humans. It is not known, though, how many non-human animals died to find out how humans die.

Flawed – Many of those who suffered were not strangers to the South African security apparatus. Whether soldiers from neighbouring countries, members of the SADF who were seen to pose a security threat, or dissidents working abroad, not all were in a direct oppositional relation to the Apartheid South Africa and thus not all stories easily fit in common post-Apartheid narratives.

Unfamiliar – Others who might be considered as victims did not suffer directly from the operations of the programme, but rather from the fallout from its revelation – for instance, the children and other relations of Project Coast staff.

Against this classification, we can further note that with few named victims or established voices post-TRC to assert the needs or interests of victims, the narrative about Project Coast (staccato and broken as it has been) has become a narrative of 'perpetrators' – and exclusively white perpetrators at that. It has been a story of their motives, intentions and capabilities, all of which have been contested. Basson as head of Project Coast has been identified as the secret keeper (and in this maintains tremendous power over those who believe they or their family members might have fallen victim to the programme). He has also garnered nearly all the attention. The reconciliation narrative of Project Coast is thus a partial story, a story of contested perpetration, revelation and reconciliation where the victims have had little voice, little identity and thus in which there could be little footing for securing forgiveness – if and when it was sought.

In the exhibition space of *Poisoned Pasts*, our response to these considerations resulted in a number of design choices. Instead of limiting ourselves to specific individuals that could be documented as having been poisoned with products proved as deriving from Project Coast, the exhibition sought to honour all those individuals who were verified, suspected or intended targets of poisoning by the agencies of the Apartheid state. In the main, the stories of those affected with this wide-ranging criterion were told in a central panel in Cluster 2 that incorporated video interviews of two affected individuals, as described earlier.

More was done though than these conventional forms of representation. Scattered throughout the exhibition were 257 small portrait panels suspended from the ceiling portraying those affected. Some of the panels take the form of a graphic silhouette, some have names, others are blank and only offer a vague description of an individual, such as 'POW' (prisoner of war). See figure 5.3.

Our decision to suspend these icono/graphic representations throughout the space was intended to work against the relegation of victimhood by framing these individuals, both known and unknown, as a vital form of absent-presence. Repeatedly encountering them within the space of the exhibition was intended to point to the viral quality of collective trauma, asserting that if such experiences are insufficiently acknowledged or understood, they will continue to produce ghosts.

Figure 5.3 Installation view of *Poisoned* Pasts at the Nelson Mandela Foundation, depicting portrait panels representing those affected by the programme, suspended in front of the 'Operations' display. They partially obscure the sightline to the 'Background' zone, the animal cages and the reconstruction of the restraint chair. (Copyright: Nelson Mandela Foundation).

The fugitive figure of the ghost entered the critical humanities with the 'spectral turn', ushered in by Jacques Derrida's *Spectres of Marx* (1993, trans. 1994).[30] Following Derrida, Peeren and del Pilar Blanco (2013) suggest that 'rather than being expelled, the ghost should remain, be lived *with*, as a conceptual metaphor signaling the ultimate disjointedness of ontology, history, inheritance, materiality and ideology. . . . As related to "the deconstructive thinking of the trace, of iterability, of prosthetic synthesis, of supplementarity, and so forth," the ghost ceases to be seen as obscurantist and becomes, instead, a figure of clarification with a specifically ethical and political potential' (emphasis in the original).[31] They further propose that the technique of 'spectrography', the writing of *and with* the spectral, offers revisionist history and geography a particular ethics that asserts relationality between people and places; it locates events and protagonists and speaks to the intersectional and transformational effects of bodies acting upon on another (2009: 483). These were values that we hoped would be communicated through our representations of those that suffered. We also intended our treatment of victims as fronting the agency (and responsibility) of those who represent or author contested pasts.

Practically, these dispersed and suspended placeholders for the absent and the missing were also intended to encourage a connection between events and people within the exhibition's narrative clusters. Wherever appropriate, affected persons were displayed in proximity to displays that describe particular circumstances. For instance, the cluster of SWAPO soldiers is installed near the image of the Namibian coastline.

Displaying those affected in this way also added a dimension to the 'sightlines' points mentioned earlier that informed the overall spatial layout. These suspended panels interrupted (and therefore animated) a focused set of thematic zones that would ordinarily be approached and read in a relatively straightforward – horizontal – way. Suspending images from the ceiling, versus displaying them on surfaces that are anchored to the floor, drew attention to spatial features such as mass (elements clustered together or widely spaced), as well as height and volume; they redirected the gaze away from a primarily horizontal reading experience and created a differentiated focal range that includes the 'surrounding' space that constituted the overall space of the exhibition.

CONCLUSION

A major source of challenge in giving accounts of the past is that the process of telling previously hidden stories will be many things to many people: resisted, traumatic, celebratory, cathartic, even banal. Some may choose to draw a line under what has passed in order to move on. For others, the past will continue to push into the present, often in discomforting ways that cannot be put aside. As Paul Williams argued in *Memorial Museums: The Global Rush to Commemorate Atrocities*, at least two critical questions arise in commemorating troubling events: 'What is the primary story to be told? Who should be authorized to tell it?'[32]

Williams identifies two popular approaches employed by those (usually institutions) with atrocity-commemoration mandates: to 'offer lessons' and to 'stage an intervention'. Conventionally, this is done by establishing the facts and visibility of the event in question, while acknowledging that current conditions represent 'a decisive break from the conditions of the [original] event'.[33] Williams suggests that a key appeal of memorial heritage projects is their dual offer of 'a concrete instance for thinking about extreme conditions and moral choices' that have defined recent history while also satisfying the lures of secrecy and danger and the enduring effects of mortality and loss. Through a layering of factual telling, emotional resonance and imaginative conjecture, memorial heritage projects allow us to 'experiment mentally with the furthest boundaries of what life can involve'.[34]

Minding such considerations, during the development of *Dis-eases of Secrecy* and *Poisoned Pasts*, a question loomed large for us: What is the (*our*) intention behind recounting the past? In other words, what do we hope histories (particularly histories of harm) can help *do*? Instruct, inform, educate? Or bear witness, honour, remember or advocate for justice? Or all or none of these?

Needless to say our individual motives, along with our respective disciplinary expertise, were somewhat different, yet complementary. They may have even changed with time; certainly the scope of initial ambitions is checked through the practicalities of exhibition making. Presenting 'old' information in new ways to new audiences – families of those affected, young scientists, the 'born-frees' for whom the TRC exists purely as a historical fact – was a common goal. Gathering visitor feedback is really the only way to construct a sense of how far we succeeded in encouraging active engagement and meaning-making among those who have visited the exhibition to date.

Dis-eases of Secrecy and *Poisoned Pasts* were not intended simply to document what took place under Project Coast. Rather we set out ways to work with what is absent, missing or contested in part to provoke questioning about the limitations of existing means to deal with oppressive pasts. Absences were not been treated as holes to be filled in, sources of frustrations or deficiencies in analysis, but rather as defining features of what was offered. Embracing the limits and recognizing the potentialities of this, we strove to promote a reconsideration of stable understandings, to free spaces for additional relevances to be brought to bear and to engage audiences in becoming curious about the relevance of the past for the future. In addition, we sought to explore ways in which the printed page and the site of an exhibition provide alternative possibilities for working with and against material forms and conventions. This was done, for instance, by presenting 'authentic' and fashioned objects together; layered and competing voices of victims, perpetrators and investigators; and non-linear narrative constructions that questioned a singular understanding.

Such strategies open themselves up to risk, including the risks of illegibility, obfuscation and frustration. But they also hold the potential for productive reconsideration of past traumas and injustice. As Maggie Nelson reminds us, with reference to critical cultural practices broadly referred to as 'tactical media', risky creative strategies are necessary to

> force the spinners and suppressors of certain facts out of the woodwork. The brutality of those facts must then hang anew, in open air, for all to see. This is not an exposure, precisely; the facts have typically been there all along. It is a means of re-attending to that which is already visible, of reconsidering that we may already know. It is, in short, a recalibration of the function of knowledge itself.[35]

Indeed, she insists, recalling Eve Kosofsky Sedgwick, such strategies may be the only way to move beyond 'the rather fixated question: Is a particular piece of knowledge true, and how can we know? to the further questions: What does knowledge *do* – the pursuit of it, the having and exposing of it, the receiving again of knowledge one already knows?'.[36]

In our contemporary moment where notions of 'truth' and the (f)actual are being increasingly destabilized, among their other aims, *Poisoned Pasts* and *The Dis-eases of Secrecy* sought to re-attend to the factual aspects of dissonant historical events that are, in Nelson's words, 'already visible' post-TRC, but which, through various processes, both willful and circumstantial, despite great civic efforts, have been re-buried, re-secreted and re-concealed. That we continue to struggle with learning lessons from history demands that the conversation be re-opened anew, in the hope that a new generation may receive it, and use it, differently.

NOTES

1 Brian Rappert and Chandre Gould, *Dis-eases of Secrecy: Tracing History, Memory and Justice* (Johannesburg: Jacana, 2017).
2 Work underpinning aspects of *The Dis-eases of Secrecy* was supported by Economic and Social Research Council (ESRC), Defence Science and Technology Laboratory (DSTL) and Arts and Humanities Research Council (AHRC) award, 'The Formation and Non-formation of Security Concerns' (ES/K011308/1), as well as a British Academy Newton Advanced Fellowship, 'Cataloguing Secrets, Transforming Justice' (SL-06825). A dataset for empirical research associated with the ESRC, DSTL and AHRC is available at https://discover.ukdataservice.ac.uk/catalogue/?sn=851599&type=Data%20catalogue. *Poisoned Pasts: Legacies of South Africa's Chemical and Biowarfare Programme* was also supported by the British Academy, an ESRC Impact Accelerator Award and the Nelson Mandela Foundation.
3 Mike Crang and Nigel Thrift, 'Introduction', in *Thinking Space*, eds. Mike Crang and Nigel Thrift (London: Routledge, 2000); Martin Jones, 'Phase Space: Geography, Relational Thinking, and Beyond', *Progress in Human Geography*, 33, no. 4 (2009), 487–506.
4 C. Gould and A. W. M. Hay, 'A Decade of Deceit: The South African Biological Weapons Programme', in *Deadly Cultures: Biological Weapons since* 1945, eds. M. Wheelis, M. Dando and L. Rosza (Cambridge, MA: Harvard University Press, 2005); C. Gould and P. Folb, 'The South African Chemical and Biological Warfare Program: An Overview', *Nonproliferation Review*, 7, no. 3 (2000), i–297.
5 Truth and Reconciliation Commission, 'Volume Two – Truth and Reconciliation Commission of South Africa Report' (Cape Town: TRC, 1998).
6 C. Gould and Peter Folb, *Project Coast: Apartheid's Chemical and Biological Warfare Programme* (Geneva: UNIDIR and Centre for Conflict Resolution, 2003).
7 W. Basson, 'Testimony at the TRC Hearing into Chemical and Biological Warfare' (Cape Town: TRC, 31 July 1998).

8 Marléne Burger and Chandré Gould, *Secret and Lies* (Cape Town: Zebra, 2002).

9 For instance, for those who did not speak or read Afrikaans, the vast majority of the testimony and documentary evidence presented in the trial was inaccessible because the trial was conducted in Afrikaans. Even for those fluent in the language, the extent and duration of the trial would be testing. Lasting over thirty months, involving 153 witnesses and relying on thousands of pages of documents that dealt with financial and drug charges beyond those associated with chemical and biological weapons, the sheer amount of information and number of contradictory claims produced created major impediments to determining what had happened. And in a bizarre twist, it became clear in 2015 after attempts by one of the authors (CG) to access the court record that the official transcript of the trial had gone missing from the court records frustrating later attempts to access it. All that remains is the judgement and daily notes taken by a Marlene Burger (a skilled court reporter) and Gould that were published on the Internet during the trial.

10 N. Knobel, 'Testimony to the Truth and Reconciliation Commission Chemical and Biological Warfare Hearings' (Cape Town: TRC, 8 July 1998).

11 Citizen, 'Chemical Warfare Papers Not Destroyed – Witness', *The Citizen*, 9 July 1998; Cape Argus. ' "Wouter's Disk" Key to Germ Warfare', *Cape Argus*, 8 July 1998.

12 Burger and Gould, *Secret and Lies*, 26.

13 Interview Project Coast scientist (requested anonymity) 20 August 2013.

14 See, for instance, O. Olwagen, 'Witch-Hunt against Basson Should End Now', *The Star*, 19 July 2012, available at https://www.iol.co.za/the-star/witch-hunt-against-basson-should-end-now-1344597

15 Laurie Nathan and Patricia Lewis, 'Preface', in *Project Coast: Apartheid's Chemical and Biological Warfare Programme*. UNIDIR/2002/12 (Geneva: UNIDIR, 2002), ix.

16 For more on this, see B. Rappert and Chandre Gould, *Dis-eases of Secrecy: Tracing History Memory and Justice* (Johannesburg: Jacana, 2017).

17 Academy of Sciences of South Africa, *The State of Biosafety and Biosecurity in South Africa* (Pretoria: ASSAf, 2015), 103–107.

18 Filippa Lentzos, ed., *Biological Threats in the 21st Century* (London: Imperial College Press, 2016), 162.

19 Reports from this dialogue series are available at https://www.nelsonmandela.org/content/page/nelson-mandela-international-dialogues1.

20 Chandre Gould, 'The Trouble with Memory Work: Reflections on the Mandela Dialogues', 27 February 2017, available at https://www.nelsonmandela.org/news/entry/reflections-from-the-2016-mandela-dialogues.

21 Following the spirit of M. Taussig, 'Viscerality, Faith, and Skepticism', in *Magic and Modernity*, eds. B. Meyer and P. Pels (Stanford, CA: Stanford University Press, 2003).

22 S. Lindqvist, *A History of Bombing* (London: Granta, 2003), i.

23 Beyond the points already made, additional intellectual origins for *Dis-eases of Secrecy* include previous efforts by Rappert to produce experimental forms of

writing intended to not only speak to how concealment and disclosure interplay within diplomatic communities but also how this necessarily affects the writing of social researchers about these communities. See B. Rappert, *Experimental Secrets: International Security, Codes, and the Future of Research* (New York: University Press of America, 2009); B. Rappert, 'Revealing and Concealing Secrets in Research: The Potential for the Absent', *Qualitative Research*, 10, no. 5 (2010), 571–588; B. Rappert 'Present Absences: Hauntings and Whirlwinds in – Graphy', *Social Epistemology*, 28, no. 1 (2014), 41–55.

24 Bennett, *op.cit.*

25 The experiment was abandoned when it proved impossible to prevent the cyanide turning the maize meal blue.

26 Wikipedia, 'Entry for Exploded-view Drawing', available at https://en.wikipedia.org/wiki/Exploded-view_drawing.

27 Ostensibly to demonstrate that despite appearing to be made of two parts (cranium and mandible), a human skull is generally accepted to consist of twenty-two separate bones (excluding ear ossicles).

28 E. Carrabine, 'Just Images: Aesthetics, Ethics and Visual Criminology', *British Journal of Criminology*, 52, no. 3 (2012), 467.

29 S. Van Rensburg, 'Testimony of at the TRC Hearing into Chemical and Biological Warfare' (Cape Town: TRC 9 June 1998).

30 J. Derrida, *Spectres of Marx: The State of the Debt, the Work of Mourning, and the New International* (New York: Routledge, 1994). See also U. Baer, *Spectral Evidence: The Photography of Trauma* (London: MIT Press, 2002) and A. F. Gordon, *Ghostly Matters: Haunting and the Sociological Imagination* (London: University of Minnesota Press, 2008).

31 E. Peeren and M. del Pilar Blanco, *The Spectralities Reader: Ghosts and Haunting in Contemporary Cultural Theory* (London: Bloomsbury, 2013), 7.

32 P. Williams, *Memorial Museums: The Global Rush to Commemorate Atrocities* (Oxford: Berg, 2007), 106.

33 Ibid., 121.

34 Ibid., 142–143.

35 M. Nelson, *The Art of Cruelty: A Reckoning* (New York: WW Norton, 2012), 161. Tactical media are associated with 'ad-busting' or 'copy-left' practices, a form of visual/media hacking that has its historical origins in the *detournement* practices of the Situationist International. Actions by the Yes Men, for example, whom Nelson is directly referencing here, have gained considerable visibility, not least because they have confronted multi-national structures and organizations that represent considerable economic power coupled with absolute ethical failure.

36 Ibid.

Chapter 6

Tear Gas Epistemology: The Himsworth Committee and Weapons as Drugs

Brian Balmer, Alex Spelling and Caitríona McLeish

International law clearly states that the use of chemical weapons (CW) as a tool of warfare is not permitted.[1] What counts as a chemical weapon is not without dispute, but the current legal definition provides some strong guidance: 'any [toxic] chemical which through its chemical action on life processes can cause death, temporary incapacitation or permanent harm to humans or animals'.[2] Use, on the other hand, is far from straightforward, as certain 'chemical weapons' have proven utilities outside of warfare but also lie firmly within the security domain, including as tools to counter social disorder. Such 'non-warfare' uses are acknowledged within the main international legal instrument that governs chemical weapons, the 1993 Chemical Weapons Convention (CWC) signed by 192 states. The treaty exempts from its scope the use of toxic chemicals for the purpose of 'law enforcement including domestic riot control' (Article 2.9.d), provided they are used in 'types and quantities' consistent with those purposes. Consequently, the same toxic chemical can be permitted in one conflict environment (law enforcement/police) but banned in another (battlefield application/military) under the CWC. This ambiguity does not appear in the earlier 1925 Geneva Protocol (GP) on the *Prohibition of the Use in War of Asphyxiating, Poisonous or Other Gases, and of Bacteriological Methods of Warfare* which also outlaws use of toxic chemicals, or to use the Protocol's words 'asphyxiating, poisonous or other gases and of all analogous liquids, materials or devices' as it only refers to the 'use in war'. As such, it was through further elaboration over the course of the twentieth century of what constitutes use, and as the focus moved from war to all hostile use, that ambiguity arose.

This chapter combines historical analysis and insights from science and technology studies (STS) to focus on the use and non-use of

2-chlorobenzalmalononitrile (CS) 'tear gas'. In particular, we analyse the work of the Himsworth Committee, established initially to look into the safety of CS tear gas after its use in Northern Ireland in 1969. Within the United Kingdom, the deliberations of this expert committee had a significant bearing on what uses of tear gas would or would not be permitted by the state. In particular, it provided a framework and language against which uses would have to be justified, and provided operational guidelines for future use. What finally was determined to be legitimate use varied across locations and in turn rested on choices, deliberations and assumptions about the epistemological and ontological dimensions of the committee's enquiry: what was known (and how) about CS gas, and what properties and effects were attributed to that gas. In this respect, the committee explicitly adopted a specific overall approach to its investigation, arguing that the issue of safety should be treated as a medical rather than military (or police) matter. In short, that CS should be assessed in terms of drug safety rather than as a weapon.

We explore some of the rationale behind, and consequences of, taking this approach. The committee, we argue, in seeking to provide both a safety assessment and guidance for future use, made a series of explicit and implicit choices about which aspects of the weapons-drugs analogy to include and exclude in their analysis. As such, we will show how the extrapolation from weapons to drugs was not automatic; it had to be made to work, and with varied success. To the extent that the analogy held, the next step for the committee was to make recommendations on future operational use. Here, we argue that the findings and recommendations of the Himsworth Committee assumed the role of a simulation, aiming to both describe and predict future use. As with all simulations, this yet again included a selective approach to what to include and exclude. Selection occurred both in the committee's attempts to shift from a weapon to drug analogy, and in attempts to extrapolate from the Himsworth findings to their recommendations for regulating future use. The result was a mix of legitimated uses at different scales: CS would be used differently in enclosed spaces than in the open; it would be used differently in Northern Ireland than Great Britain. The recommendations, however, floundered at several points, such as when the police came to deploy tear gas in Scotland, exposing the fragility of both the 'weapons as drugs' analogy and the supposed precision that carried into recommendations about tear gas safety in situations of civil disturbance.

UNPACKING THE USE OF 'USE'

The question 'What is the use of a chemical weapon?' can be responded to along different lines: Is it a useful or useless weapon? To what uses has or can it be put? For whom is it useful? Under what circumstances would a chemical weapon be permitted? Under what circumstances would a chemical weapon be effective? Put bluntly, 'use' is not a straightforward term, particularly because some of these questions are descriptive, asking what are the properties or effects of a weapon, and others are normative, asking when or if we ought to employ the weapon. Moreover, as the editors point out in the introduction to this volume, even labelling a chemical agent as a 'weapon' is an evaluative act that evokes connotations about appropriate and inappropriate use.

Scholarship on the history of chemical and biological weapons (CBW) has repeatedly shown how judgements about utility are related to the context of use. So, concluding an edited collection on the history of twentieth century biological warfare (BW) programmes, Dando and colleagues noted that the changing perceived utility of biological weapons (BW) was always a relative judgement – primarily made by comparison with the changing perceived utility of nuclear weapons.[3] Martin has recently argued that the U.S. decision to use anti-plant and irritant agents in the Vietnam War was a complex mix of considerations.[4] She notes that both pragmatic military utility and normative/ethical deliberation were involved – but primarily, she argues, the former dominated the decision-making process. More generally, Robinson argued some time ago that, historically, chemical weapons have often been of most use by technologically superior northern countries (or southern countries able to appropriate the technology) against less prepared and protected southern countries.[5] This axis, he argued, might be reversed in a *David v. Goliath* desperate situation such as in order to panic urban civilians, in terrorism, or in conjunction with chemical weapons. He concludes: 'Poison gas is, therefore, peculiar in that its value as a military weapon seems to be far surpassed by its value as a political weapon'.

These observations about the relational nature of 'use' resonate with scholarship within the sociology of technology. This work has demonstrated how seemingly fixed or self-evident properties of weapons – such as their utility, effects and efficacy – are context dependent.[6] For instance, Rappert has undertaken extensive empirical work to demonstrate how ambiguities around so-called non-lethal weapons (NLWs) are constructed.[7] Recent approaches within STS have developed these and related ideas to establish how the 'social' and 'technical' do not stand as separate domains with one impacting on the other. Instead, empirical and theoretical work within the 'co-productionist idiom' has shown how social and natural orders are produced simultaneously.[8] So, for example, if poison is designated as a sabotage weapon, it constructs an identity for *both* the chemical and its user. Likewise, in the nineteenth century when some criminologists designated poison as a 'women's weapon', because it was deemed a weapon of the weak, this designation was simultaneously constructing ideas about the properties of both the weapon and of women.[9]

These approaches to thinking about technology allow us to unpack 'use' in relation to tear gases. If the utility of a chemical agent is not simply given, it becomes fruitful to undertake further detailed empirical investigation of just how and why they have come to be regarded (or rejected) as permissible and 'the right tools for the job'.[10]

We also suggest that the characterization of certain chemical agents is closely inter-twined with different forms of epistemic authority, which are 'licenced' to make particular claims about those weapons. At different times, experts such as scientists, lawyers, diplomats, civil servants and politicians, military commanders and the police have all made claims, based on their own domains of expertise, about what particular chemical agents are (e.g. lethal, non-lethal, humane, strategic and tactical) and what these properties imply about what ought to be done with that chemical agent. STS and history of science have long recognized how science intersects with different forms of expertise such as lay knowledge, legal or military expert authority. Here, we argue that in our case study, tear gas had a large number of 'epistemic constituencies' speaking on its behalf.[11] Our focus is on *what* such expert constituencies know, and its subsequent effects in policy making and implementation, but equally – writing as sociologists and historians of knowledge – on the processes involved in *how* they come to know it.

THE HIMSWORTH COMMITTEE AND WEAPONS AS DRUGS

In August 1969, steadily building tensions and conflict between Catholic and Protestant communities in Northern Ireland erupted in the city of Londonderry.[12] In what came to be called the 'Battle of the Bogside', several days of rioting took place in the wake of an Apprentice Boys march on 12 August.[13] The Royal Ulster Constabulary (RUC) resorted to using the chemical agent CS, fired as cartridges and thrown as grenades, during the following two days of disturbance. CS, commonly known as tear gas, belongs to a class of chemical weapons known as lachrymators, so called because the most obvious symptom of exposure is the flow of tears accompanied by a range of other symptoms such as involuntary eye closure and pain in the eyes and throat.[14] Once removed from exposure, these main effects tend to be quite short-lived, hence its apparent suitability for use in controlling civil disturbance.

Within a month after this first use of CS in Northern Ireland, Home Secretary James Callaghan established a small committee to carry out 'an investigation of the evidence relating to the lasting medical effect, if any, of the CS control agent upon persons exposed' during the rioting.[15] The committee was chaired by Sir Harold Himsworth, a clinical researcher who had recently retired as the secretary of the UK Medical Research Council (MRC), the quasi-governmental body responsible for distributing public research funds in the area of biomedical research.

Himsworth had safeguarded a strong tradition within the MRC to maintain the autonomy of researchers from government (and MRC) steering, insisting 'all that organisation can do is to choose the right man [*sic*] as leader, equip him to the best of its ability with men and materials, and trust to his judgment'.[16] Moreover, with regard to providing policy advice to government, he had previously made the position of the MRC clear, affirming that 'it is only insofar as a research council can maintain a reputation for scientifically objective advice and policy that it can serve either society or the world of research'.[17] Himsworth, therefore, was an ideal choice in terms of being both an establishment figure but also one who might be seen to be independent in the politically charged context of Northern Ireland. The other two members of the committee were Professor Anthony Dorhorst, professor of medicine at St Georges Hospital Medical School, and Professor Robert Thompson, Courtald professor of biochemistry at the Middlesex Hospital Medical School also chair of the Chemical Defence Advisory Board (CDAB). CDAB provided technical advice primarily to the UK's main site for CW research, the Chemical Defence Experimental Establishment (CDEE) at Porton Down, Wiltshire.[18] As the Himsworth Committee's first report explained, Thompson clearly had some knowledge about CS – while Himsworth did not – but his Porton association inevitably pulled against the symbolic independence lent by appointing Himsworth as chair.

In compiling their initial report, the committee members visited Northern Ireland and the CDEE. This first report confined its coverage to the situation in Derry but ended with a more general statement from the members that was to underpin their approach in a second, more general and lengthier study of CS gas.

> Finally, we would like to express an opinion to which we have come in respect of chemical agents that might be used for civil purposes. In our opinion, from the point of view from which the effects of any such agent should be studied should be more akin to that from which we regard the effects of a drug than to that from which we might regard a weapon.[19]

This 'weapons as drugs' approach could, indeed, be regarded as one of the lasting legacies of Himsworth. Just as this chapter was being completed, Feigenbaum published a wide-ranging history of tear gas, including an analysis of CS in Northern Ireland and the work of the Himsworth Committee.[20] Feigenbaum also notes the weapons as drugs approach and furthermore emphasizes how the committee members operated with a hierarchy of evidence that gave credibility to the results of technical, laboratory-based studies and generally dismissed first-hand eye witness accounts of harm from tear gas.

Several more recent discussions of tear gas draw attention to the legacy of the 'weapons as drugs' framing of Himsworth's fuller investigation.[21] In this vein, Rappert's study of more recent and potent CS *sprays* is critical about the processes of safety assessment. He argues, based on interviews and documentary research, that official claims that sprays 'have been or are regulated in a manner similar to that of a pharmaceutical drug are highly questionable'.[22] Wright goes further in his interpretation, claiming that Himsworth's recommendation to treat weapons as drugs was due to inadequate data on the agent and so was part of using science to make CS gas '*appear* rather than to *be* safe'.[23] He continues that, as a consequence, CS and other such control technologies reconfigure the rules and norms about what the state deems acceptable in countering 'political opposition or unruly citizens'. This point, about the apparent charade of science, is echoed by Feigenbaum, who notes that the 'weapons as drugs' approach had already been adopted at Edgewood Arsenal, the main chemical weapons production site in the United States, as partly a 'public relations ploy' to place CS in a separate regulatory category to small arms and chemical weapons.[24] Moving beyond CS and examining the immediate legacy of Himsworth, in his meticulously researched history of chemical and biological disarmament during this period, Walker points out that because of the 'weapons as drugs' stipulation, the same approach was applied a few years later to considerations about authorizing use of more potent agent dibenzoxazepine (CR). Noting a certain nervousness in these discussions, he concludes 'despite the very clear and conclusive safety studies, Ministers perhaps only subconsciously, may have harboured lingering doubts over [CR] safety'.[25]

All of these recent analyses consider the 'weapons as drugs' decision after the publication of the Himsworth reports. In this respect, the decision itself has outlived the details of the decision-making and safety assessment process. What follows, as mentioned at the outset of this chapter, is a re-visiting of the second Himsworth report, reading it in conjunction with the first report and through the perspective of the history and sociology of knowledge. It is interesting, in this respect, to ask questions about the Himsworth Committee in terms of what might be called 'tear gas epistemology' and the problematic nature of 'use': How did the 'weapons as drugs' framing operate? How robust was it and what rationales underpinned it? As we now demonstrate, the framing attempted to smooth over a number of dissimilarities between drugs and weapons safety, and at once widened the enquiry in a positive way and narrowed it in a negative way.

THE SECOND HIMSWORTH REPORT

The committee was expanded in terms of membership and expertise, and its second report was published on 15 September 1971. Expanding on the logic of the 'weapons as drugs' approach, this report argued that drug safety had both a general and a particular element to it.[26] At one level, there was general knowledge about the safety and toxicity of a drug. Such knowledge then needed, they explained, to be applied in a clinical context of balancing the severity of a patient's illness against the risks of taking any particular dose of the drug. Shortly afterwards, the report noted that an ideal drug would be harmless regardless of dosage but, because such ideal drugs were rare or non-existent, that in reality, there should be some safety margin between the effective and harmful dose. The same principle, they argued, would apply when considering an riot control agent (RCA). Himsworth was thus forging an overt 'translation' – a term used in the report – of the CS problem from one primarily involving military expertise to one primarily involving medical expertise. Indeed, Himsworth points out that in the case of decisions about administering a toxic drug for treating cancer and for a 'relatively trivial' disease, 'these are both medical considerations and the responsibility for striking the balance between them is purely a professional matter'.[27]

Having made a case for appropriating the CS problem to medical expertise, the committee recognized that the analogy could not undergo a smooth and seamless transition. In this respect, they noted that in an anti-riot situation, some decisions would be 'purely medical', in particular the health risks of the agent, while 'others raise questions of social policy and must be assessed on political as well as medical grounds'.[28] The committee also highlighted specific points of friction that had arisen when they tried to translate the anti-riot setting into the drug safety setting. While drugs, they reported, are generally administered on an individual basis in precise, calculated doses, tear gas is administered in the form of a smoke to groups, where any particular individual would be exposed to a highly circumstance-dependent amount of the agent.

This slippage from a controlled to uncontrolled situation continued. Those exposed could include both rioters and people in the vicinity including children, elderly people and pregnant women. Finally, the committee pointed out that with sparse evidence of the effects on humans, they needed to turn to animal experimentation data. These extrapolations – individual to population; healthy adult to more vulnerable; animal to human – were underpinned by a leap in spatial and socio-technical imaginaries.[29] Despite acknowledging some points of friction, by declaring that the committee would adopt the analogy, the hostile and confrontational riot, with its accompanying confusion, contingencies and unpredictability, was reconfigured in the report as an orderly place where prediction, control and relative calm prevail. Likewise, the largely consensual situation of receiving a drug was linked to non-consensual crowd control. Arguably, this move also created an elision throughout the report between non-lethal and non-violent, because use of CS gas to quell civil disobedience still remains an act of violence.

The remainder of the second Himsworth Committee report considered experimental evidence on the safety of CS. It included an estimation of the safety margin between effective and dangerous doses, and the mode of action of inhaled CS. The report then considered CS exposure in healthy adults, groups such as the elderly or pregnant and those with pre-existing conditions such as respiratory diseases. It also contained a lengthy section on operational matters concerning the circumstances of CS use. As mentioned, the committee combined evidence from animal experiments with reports from Northern Ireland and elsewhere, including evidence from CS use in the 1968 Paris riots and Vietnam war. The report acknowledged uncertainties in the data, for example, placing a number of caveats around its conclusions on chronic effects such as carcinogenicity.

It also discounted evidence, in at least one place, on the grounds of epistemic constituency. In the course of a discussion of whether CS had rendered people unconscious in Londonderry, the committee acknowledged a visit to Northern Ireland by members of the British Society for Social Responsibility in Science (BSSRS) comprising 'sociologists . . . and medical men'. The medical section of the report concurred with the Himsworth Committee finding: there had been no cases of loss of consciousness. On the other hand, the sociologists recorded people witnessing other people being rendered unconscious but, with blatant dismissal of sociological testimony, the report concluded that 'in the light of the fact that no such cases were reported by the medical section of this [BSSRS] enquiry we feel that this statement cannot be accepted as having been substantiated'.

The report's conclusions have been widely discussed, both at the time and in the more recent literature cited earlier. The main claim Himsworth made was that it was 'highly improbable' that CS used to quell civil unrest would result in serious damage or fatality. The committee warned of some possibilities of minor injuries, noted that CS might exacerbate bronchitis or trigger asthma attacks in some but not all cases, flagged uncertainty over whether the agent was carcinogenic, and urged caution in using the agent in confined spaces, where the dosage would be higher.

FRAMING AS WIDENING AND NARROWING ENQUIRY

Despite criticism of the findings of the report, it appears that the 'weapons as drugs' framing of the investigation had consequences that both opened up and closed down further debate. In terms of opening up, it meant that Himsworth widened out the scope of who should be considered in the assessment of effects. As the first Himsworth report concluded, by treating CS safety as akin to drug safety: 'the effects of any such agent should be appropriately investigated, not only in respect of healthy persons against whom it might be directed, but also in relation to the possible effects on the young, the elderly and those with impaired health, who might be inadvertently exposed to the agent in question'.[30]

On the other hand, 'weapons as drugs' established a benchmark assumption that the status quo would be continued use of CS gas unless evidence of harm emerged from the enquiry. An alternative could, in the face of the uncertainties acknowledged in the assessment of CS, have been for the committee to call for a moratorium until further evidence emerged. Here, the significant point is not so much that a case could be made for either continued use or a moratorium, but that an assumption was made and sustained that the demonstrable utility of CS in Northern Ireland had created a default of continued use (with, presumably, little need for further monitoring). Furthermore, the particular formulation of the 'weapons as drugs' approach created an artificial division between a group of rioters and a group of non-participants, inadvertently exposed because of an accident of geography. This split meant that any division between violent and non-violent protesters within a crowd became elided throughout the Himsworth report. Everyone participating in protest was tacitly assumed to be a legitimate target.

More generally, the committee's distancing of medical epistemic expertise from 'social policy decisions' created an artificial separation between natural and social orders.[31] Their treatment of operational matters later in the second report demonstrates how their 'technical' assessment could not avoid making judgements about social order. In discussing the use of CS in the more risky environment of confined spaces, they noted: 'We have found no reason to believe that CS gas had been used lightly by the armed forces or the police. Nevertheless in the rapidly moving circumstances of civil disturbances there is always the risk that the judgement of the individual directly involved may be overhasty', thus acknowledging that having rules of conduct does not ensure their enforcement. Yet, very shortly after, the report points out that strict instructions were given in police training manuals about dealing with panic and the possibility of injury after people are exposed to tear gas. The report continued, however, that 'in assessing the possible dangers from using CS gas in operational circumstances, we have taken it that these instructions will be enforced and that, in consequence, we are at liberty to confine our attention to the toxicological aspects of this case'. This oscillation between deferring to or challenging the authority of the police training manual and practical expertise shows how the committee could not avoid making judgements about how CS gas might or might not be deployed. Returning to our earlier theoretical discussion, this discursive manoeuvring shows how the safety of tear gas and its circumstances of use (i.e. whether or not instructions would be obeyed) were co-produced in the report. Additionally, labelling CS as safe, an ostensibly scientific decision, would possibly contribute to more widespread use. So, even if this last point is conjectural, it further demonstrates that scientific and social considerations could not be isolated from each other.

HIMSWORTH AS SIMULATION

We have already noted that the transposition, from military to medical, of epistemic constituencies that were delegated authority to ascribe properties to CS was not seamless. In multiple small ways, some acknowledged by the Himsworth committee, the chemical-agent-in-a-riot resisted being 'sedated' into the ambit of a medical framing. Although there is a large and varied literature on simulation, we argue here that it is fruitful, from an STS perspective, to view the entire Himsworth report as a 'simulation' for two reasons. First, because the investigation had to selectively include and exclude aspects of the actual circumstances of use in order to construe weapons as drugs in the manner it did. Second, because the findings and their interpretation based on this analogy informed a lengthy section in the second Himsworth report that made predictions about different conditions in which tear gas might be deployed and offered recommendations on future operational use.

In his wide-ranging discussion of the relationship between war and video gaming, Crogan defines simulation as: 'a process by which a phenomenon is representatively modelled by another phenomenon. The process involves a selective reduction in the representative model of the complexity of elements composing the simulated phenomenon'.[32] That this reduction is a social process is highlighted in Ghamari-Tabrizi's close study of RAND's Cold War role-play and gaming simulations of nuclear war.[33] In these situations, where the simulation was of something that had never occurred, the designers frequently found themselves having to make a spectrum of judgements about which elements of the simulation had to be authentic and which were trivial. So, clearly the precise warning time available before a nuclear attack was highly pertinent to the outcome of the role-play, whereas the relevance of faithfully copying most other features in the simulation remained open to judgement and discussion.

The Himsworth committee second report and associated evidence effectively performed similar selective operations. As discussed, they took notice of only certain aspects of the 'weapons as drugs' analogy and smoothed over differences, in order to use observations from the Northern Irish and other geopolitical events and combine them with experimental data to create predictions of the effects of CS in any situation. In the process, two elements of Crogan's characterization of simulation are also evident with Himsworth: pre-emption and the blurring of the real and simulated.

First, as Crogan documents, the history of computer simulation – from early cybernetic attempts to predict flight paths of enemy aircrafts, through flight simulators, to contemporary attempts to simulate the battlefield – has been simultaneously about copying and controlling what is being simulated. He emphasizes the novelty of computer simulation in the automation of war, arguing 'if combat means a contest between two or more opposing people, then these simulational technics effectively anticipate it and pre-empt its occurrence'.[34] We should not be distracted by the emphasis Crogan places on new technology – attempts to automate war are evident in much older historical attempts to write instructions and to drill military personnel so that they act in a predictable and controllable manner.[35]

Accepting this point reinforces our argument that the second Himsworth report be construed as a simulation designed to move from the specifics of Derry streets and to predict and control universally, by instructing how to use CS 'safely'. Moreover, as Rappert has observed, the use of so-called NLWs – simply to keep them non-lethal – involves a far more intricate choreography around instructions and rules than weapons used with lethal intent.[36] Hence, the Himsworth report, with its instructions about safe operational use in the open, in confined spaces, in the vicinity of people with pre-existing conditions and so on, amounted to far more than just a scientific investigation into the safety of a chemical agent. Likewise, this observation reinforces the point that the reductions and parallels drawn by Himsworth between drugs and weapons, riots and clinics, were not straightforward and obvious, but instead were the result of a series of choices and implicit rankings of what was deemed more or less pertinent to the veracity of the analogy.

As a second element of simulation, following Baudrillard, Crogan argues that the simulation becomes seamlessly 'real' (hyper-real) alongside that which it simulates. It does this by re-inserting itself back into the world as the 'real' reference point for acting and thinking in relation to a phenomenon. Again, the move between the first report with its local focus and the second report that aimed for generalization exemplifies this relationship between the real and simulated. The resistances encountered, and partially acknowledged by the Himsworth committee, in the translation from military to medical expertise have already illustrated how the process of 'selective reduction' from Crogan's definition of simulation does not happen without work to make it happen. Indeed, the attempt to create such a reference point was far from unproblematic and as we discuss next, the hyper-real simulation acted as the 'real' reference point for tear gas only in certain respects. The second Himsworth report ostensibly stood in to represent (and control) all non-military situations involving CS use, but, on the other hand, its authority to be extrapolated to other chemical agents was to be resisted.

WEAPONS AS DRUGS: SOME TRANSLATIONS

CR in Northern Ireland

Events following the Himsworth committee report demonstrate a certain fragility to the 'weapons as drugs' framing. Prior to its publication, a meeting was held between Home Office and Ministry of Defence (MoD) staff to discuss the findings of the report. At the meeting, and in several related documents, civil servants claimed that the origin of the 'weapons as drugs' doctrine was not clear. Himsworth had, they pointed out, either assumed that the acceptance of the first report meant that the approach had been approved or possibly it had been agreed verbally with Callaghan (now no longer Home Secretary).[37] The Chief Scientist of the Home Office then expressed doubts about the approach:

> Exhaustive tests required experiments on many animals over a period of time, and this sort of examination was not applied to everyday substances, only to those about which there were suspicions.

Another attendee pointed out that Himsworth had used the term 'more akin to' drugs.

Issues over the ambiguity of the term returned in 1973 as the government considered the possibility of authorizing a different agent, CR. Their plan was to make it available at the Maze prison, where paramilitaries in Northern Ireland were being detained, in case of a mass escape attempt. CR, a far more painful and irritating agent than CS, had been developed and tested at Porton Down since 1961.[38] Safety tests had been performed, and some already published, about CR but doubts had been raised in the Foreign and Commonwealth Office (FCO) over whether these met the 'weapons as drugs' standard of the Himsworth report. A written discussion of the issue within the MoD claimed that the FCO might be 'overstating the Himsworth requirements' and that a different phrase had been previously used in FCO correspondence: 'with all the vigour that would be applied in the case of a new drug'.[39] The implication was that although there had been a range of safety tests performed on CR:

> One could not necessarily expect that the detail and range covered by Himsworth, which included procedures and an analysis of CS used in operational conditions, and which in [any] case arose from unique circumstances, should be repeated for each ensuing agent.

At the same time, a letter on the CR matter sent from David Young (Private Secretary to the Minister of State for Defence) to Michael Alexander (Assistant Private Secretary to the Secretary of State in the FCO) also loosened the terms of the Himsworth approach, stating 'it does not say that the procedures should be identical in all respects with those pertaining to a new drug and in our view we have satisfied the proposed standard'.[40]

A week later, a MoD document was circulated drawing attention to the parts of the Himsworth report likely to be relevant to discussions around CR safety. Here, the author commented that 'more akin to the effects of a new drug than to that which we might regard as a weapon' probably meant that Himsworth had considered that there was a spectrum from drugs to conventional weapons, with chemical agent testing falling 'towards the drug end of the spectrum, but [would] not necessarily reach it'.[41] At this point, it appears the 'weapons as drugs' approach had dissolved into what could loosely be called a 'drugs as almost weapons' interpretation.

Tying together this slippage and the earlier discussion of simulation highlights the contrast introduced in the previous section. Himsworth was accepted as a model for other non-military use of CS, yet it was not to be copied in other respects when extending the report to a different chemical agent. The general thrust of the various qualifications to the weapons as drugs approach sought to depict Himsworth's work on CS as an isolated case rather than a paradigm. In this instance, the agent under consideration was CR, but the slippage implies that if Himsworth's analogy did not apply here, then it would not do so for any other future cases. Another reason for drawing attention to this shift in definitions is more theoretical. Within STS a number of studies have shown how controversial issues become framed in ways that foreground and background particular issues.[42] What has been neglected is to acknowledge, as demonstrated by the slippage in the Himsworth 'weapons as drugs' frame, that as they travel, temporally or spatially, frames can be more or less robust. Moreover, as discussed earlier, the general 'weapons as drugs' framing has lasted as a legacy of Himsworth and enters more contemporary debate; yet at a finer-grained level, the actual meaning of the frame and what it entailed can be seen to have been challenged and modified relatively quickly.

CS Grenades in Scotland

Himsworth had, as already mentioned, raised the issue of CS use in enclosed spaces. Without ventilation to disperse the CS cloud, the committee considered that if someone was unable to exit a room, he or she might be exposed to dangerous levels of the agent. The committee had therefore recommended that a smaller cartridge be developed and adopted for use in circumstances such as trying to remove an 'armed criminal or mentally deranged person' from a room.[43] Prior to the public release of the second Himsworth report, the Home Office moved on the matter and advised police Chief Constables that CS cartridges and grenades should not be used in rooms smaller than 50 and 100 cubic metres.[44]

These recommendations were brought into sharp focus shortly afterwards. On 19 October 1969, the police used CS in Kirkcaldy, Scotland, to force an armed man – who had already fired shots – to leave his house.[45] Two CS grenades had been used and, in a letter from the Scottish Home and Health Department to the Home Office, it was pointed out that the living room and bedroom where the grenades detonated were 'very much smaller than the proposed safety limit of 100 cubic metres'.[46] The letter continued that 'had the police been notified of the safety limit they would have been faced with a difficult decision and had CS smoke not been used, there would have been a very high risk of police officers being wounded or killed'. Chief constables, the letter suggested, should be given leeway to balance the risk of injury to an armed criminal against the risk of injury or death to police or the public. A later note pointed out that, in relation to the Home Office guidance, in a situation on the ground, 'the police are not likely to be able to have exact measurements of the places in which armed criminals are besieged'.[47]

While relatively minor, this incident and the response to it with respect to the safety of tear gas demonstrate how the Himsworth 'weapons as drugs' approach, and particularly the risk modified by severity of disease calculus of medical expertise outlined earlier attempted to translate back into everyday policing. Put bluntly, it shows how there was no smooth re-insertion along the lines of Baudrillard's notion of simulation: the rules required interpretation in the light of both the practicalities of guessing the size of the room and the need to balance the safety of the target against that of the police officers (or other members of the public). So, in this instance, the letter from the Scottish Home and Health Department and subsequent note recognized a role for police experience and expertise, which extended beyond the science of the Himsworth report, in deciding whether or not tear gas was safe or not to use.

CS in England and Wales

Enmeshed in the contrasting geographies of 'safe' open air use of CS and 'dangerous' use in confined spaces was a more obvious geographical divide. The Scottish police had to answer the question about where tear gas could be used in terms of the estimated size of a confined space. Behind these considerations, lay broader political decisions and national differences about where CS would be employed. CS use in Northern Ireland had crossed a threshold because the British government had previously restricted it to colonial use.[48] It soon became apparent that Northern Ireland, nonetheless, was to be treated as exceptional territory with respect to CS. Accompanying the public release of the second Himsworth report was a list of potential questions and answers about its contents. One question specifically noted that use would not be the same in England and Wales:

Q8: Will CS gas continue to be used in (a) Northern Ireland and (b) by the Police Forces in England and Wales?
A8: Yes, in the appropriate circumstances.
Q9: What are the circumstances in which it would be used in England and Wales?
A9: Chief officers of police have agreed only to use CS for dealing with an armed criminal or violently insane person besieged in a building or similar place from which he cannot otherwise be dislodged without danger or loss of life.[49]

The incident in Kirkaldy suggests that similar rules for England and Wales had been adopted in Scotland. No elaboration was provided here about the 'appropriate circumstances' in Northern Ireland. However, a later set of notes prepared as background for any supplementary questions in Parliament on the use of CS gas reported that 'an officer of the RUC not below the rank of district inspector may order the use of CS where it is the most humane and effective way of dealing with disorder'.[50] An attached background note added that 'there is no intention that CS should be used in Great Britain for dealing with public disturbances'. Returning to our earlier discussion of use, it is clear that who counted as the legitimate target of tear gas and what counted as legitimate deployment reflected more entrenched attitudes to the differences between Northern Ireland and mainland Britain. In this case and the previous case involving the police, wider considerations were re-introduced at different scales to create rules and decisions about different uses of tear gas in different places.

GEOPOLITICS, TEAR GAS USE AND PROHIBITED USE

The Himsworth recommendations on tear gas also intersected with discussion at the geopolitical scale, particularly with regard to an ongoing controversy over the legality of tear gas. As mentioned at the starting of this chapter, tear gas is ambiguous insofar as it is banned for use in war by the 1925 GP. Domestic riot control is simply not mentioned in the treaty. The specific wording is a prohibition in war of 'asphyxiating, poisonous or other gases'. In February 1970, in the midst of the Himsworth Committee deliberations, the British government announced that they interpreted international law to mean that the GP did not cover CS. It was therefore regarded as being permitted for use in both domestic and international conflicts. This announcement contradicted a statement from 1930 that the British government interpreted the Protocol to mean tear gases were banned in war. The convoluted discussions that led to this decision, and the subsequent controversy it generated within government and in public, have been untangled and written about in some detail by one of the authors of this chapter.[51] Here, we draw attention to its relevance to Himsworth and also how some aspects of the wider discussion before and after the February 1970 announcement drew on different types of epistemic authority to assign properties to CS.

The main line of division in this discussion ran between ministries. Staff in the FCO wanted to reaffirm the 1930 decision; within the MoD, it was thought that CS should be excluded. The MoD case, in essence, was that it would be odd for a weapon currently being deployed in Northern Ireland to be banned in more geographically distant theatres of war. Moreover, in an appeal to scientific expertise, they argued that CS was not toxic and certainly far less harmful than its predecessor chloroacetophenone (CN). The FCO officials turned in a different direction and consulted the legal expertise of the attorney general. Toxicity was simply side-stepped; indeed the Protocol makes no mention of toxicity. In the attorney general's judgement, the issue turned on the phrase 'other gases' in the wording of the GP, pointing to its redundancy unless it meant all gases that were neither asphyxiating nor poisonous. Hence, it would include CS. The MoD response was to shift ground and argue that CS was not a gas but actually a smoke. This claim might have been derived from the ongoing work of the Himsworth Committee, which had already reported and was preparing its second report. In the second report, however, the committee clearly states:

> Although CS is commonly represented as being a gas, it is in fact a smoke or fog of suspended droplets or particles that issues from these devices. From a medical point of view this distinction is important.[52]

So, the basis of the MoD argument involved an appeal to scientific and medical expertise about toxicity and whether CS was a smoke or gas. For the FCO and the attorney general, the question was not about the epistemic authority of science, but the legal expertise needed to interpret a treaty.

As corollary to this invocation of different types of expertise to decide on the status and identity of tear gas, and its terms of use, the two ministries also crossed swords over diplomatic expertise. The FCO had been clear that the decision about the GP would put the United Kingdom in an awkward position during the ongoing negotiations in Geneva to establish a Biological Weapons Convention (BWC). Being seen to go back on its 1930 decision over the Protocol, which covered both chemical and biological (bacteriological) weapons, would, they contended, lose them a lot of political goodwill in Geneva. By way of response, the MoD deigned to give the FCO a lesson in diplomacy arguing 'international goodwill is not a quantifiable quantity', and that they should concentrate not on such an ephemeral notion but instead seek calculated deals.[53]

Himsworth was tied into this broader political discussion about the international legality of tear gas. Much internal discussion took place on the matter, and civil servants prepared quite elusive answers about the second report findings and the GP should the topic be raised. In attempting to resolve the matter, government officials had turned to quite different sources of epistemic authority to bolster their respective arguments.

CONCLUSION

Our close look into the workings of the Himsworth Committee has shown how different expertises were mobilized and appropriated in decisions around the use of CS. We have argued that the shift from CS safety as a military-police matter (concerned with placing limits on violence) to a medical matter (analogous to administering a drug) was not a frictionless comparison. As we described, the committee members recognized that this transposition involved resistance, as some elements of the two situations were clearly different. Other differences, such as the degree of consent in military and medical situations, remained unexpressed but were nonetheless relevant to the deliberations of the committee. Weapons as drugs widened the scope of the enquiry to include groups that would have been invisible under a military framing but equally narrowed the enquiry by importing assumptions about the state in the role of physician rather than in its traditional role as the legitimate user of violence.[54] Furthermore, we argued, the weapons as drugs framing – because it selectively included and excluded aspects of the analogy – facilitated an enquiry that was more akin to a simulation exercise. Such simulation embodied aspirations for both replication of the present and control of the future, as the second Himsworth report claimed a generalizability that ostensibly outlived the situation on the streets of Londonderry.

There is also an extent to which the work of the Himsworth Committee exemplifies how the geography of knowledge is enacted. The reports reduced the safety of tear gas to one of scales: tear gas was judged to be safe to use in the open; safe only within specified constraints in contained spaces and not safe within the smallest of spaces. The Kirkaldy incident demonstrates how the police had to re-interpret Himsworth's recommendations based on their experience and knowledge of dealing with dangerous situations. Then, at the national scale, officials clearly made additional judgements about the different permitted uses of CS in Northern Ireland and Great Britain. From a geopolitical perspective, the different views propounded by the British MoD and FCO over the place of CS gas in the GP were, at the same time, different opinions about whether the boundary between the domestic and foreign (battlefield) marked a boundary between legitimate and illegal uses of tear gas. These scalar differences suggest that the geography of knowledge is far more than a statement that different people claim to know different things in different places. Instead, it suggests that scale helps in constructing what is known about tear gas safety and consequent variations in how and where it is legitimate to use.

Looking retrospectively at how the Himsworth Committee arrived at its decisions about safety and recommendations about permitted use illustrates how, in the United Kingdom at least, tear gas was constructed as an agent for domestic riot control in Northern Ireland. The legacy of the Himsworth Committee, we have observed, has been to treat CS agent safety as a drug safety issue. This framing implicitly naturalized the distinction between lethal and NLWs: tear gases, apparently, are 'by their nature' non-lethal. Our chapter has shown, on the contrary, that this distinction, which emerges from discussions of use and safety such as contained in the Himsworth reports, is simultaneously a social and technical achievement. It certainly involved technical and scientific assessment of evidence but this was far from being divorced from assumptions about the context and circumstances of use. This larger claim will come as no surprise to scholars familiar with developments in STS, although our case study sheds light on this process in the context of the politically significant question of how weapons come to be used.

Taking this point forward, the blurred boundaries around utility also raise wider questions for future research about the appropriateness of particular conceptualizations around which chemical weapons governance structures have been developed, such as binary conceptualizations (lethal/non-lethal; toxic/not toxic; covert/overt; small scale/large scale; domestic/international; civil/military) and quasi-scientific conceptualizations such as 'life processes' – a phrase used within the CWC to define what is/is not to be regarded as a chemical weapon. If the case of the Himsworth Committee sensitizes us to anything, we should expect these categories to be far from self-evident and instead should invite close, careful analysis of how they emerge, stabilize, change and dissolve both temporally and geographically.

NOTES

1 This work was supported by Arts and Humanities Research Council (AHRC) grant Award Number: AH/K003496/1, Understanding Biological Disarmament: The Historical Context of the Biological Weapons Convention, and Spanish Ministry of Economy and Competitiveness Research Project (FEDER) FFI2015-64529-P on the boundaries of science and politics. We are grateful to Charlie Norell for additional research assistance with the archival work required to write

this chapter. Brian Balmer is grateful for exceptionally helpful feedback on an earlier draft from Jon Agar, Sarah Edwards, Amparo Gómez, Phyllis Illari, Frank James, Brian Rappert and Simon Werrett.

2 '1993 Chemical Weapons Convention, Article II', OPCW, available at https://www.opcw.org/chemical-weapons-convention/articles/article-ii-definitions-and-criteria/.

3 Malcolm Dando, Graham Pearson, Lajos Rózsa, Julian Perry Robinson and Mark Wheelis, 'Analysis and Implications', in *Deadly Cultures: Biological Weapons since 1945*, eds. M. Mark Wheelis, Lajos Rózsa and Malcolm Dando (Cambridge, MA: Harvard University Press, 2006), 355–373.

4 Susan B. Martin, 'Norms, Military Utility, and the Use/Non-Use of Weapons: The Case of Anti-Plant and Irritant Agents in the Vietnam War', *Journal of Strategic Studies*, 39, no. 3 (2016), 321–364.

5 Julian Perry Robinson, 'Chemical-Weapons Proliferation in the Middle-East', in *Non-Conventional Weapons Proliferation in the Middle East: Tackling the Spread of Nuclear, Chemical and Biological Capabilities*, eds. Efraim Karsh, Martin S. Navias and Philip Sabin (Oxford: Clarendon Press, 1993), 69–98.

6 For reviews, see Brian Rappert, Brian Balmer and John Stone, 'Science, Technology and the Military', in *The Handbook of Science and Technology Studies*, 3rd ed., eds. Edward Hackett, Olga Amsterdamska, Michael E. Lynch and Judy Wajcman (Cambridge, MA: MIT Press, 2007), 719–740; Kathleen Vogel, Brian Balmer, Sam Weiss Evans, Inga Kroener, Miwao Matsumoto and Brian Rappert, 'Knowledge and Security' in *The Handbook of Science and Technology Studies*, 4th ed., eds. Ulrike Felt, Rayvon Fouché, Clark A. Miller and Laurel Smith-Doerr (Cambridge, MA: MIT Press, 2017), 973–1002.

7 Brian Rappert, 'The Distribution and Resolution of the Ambiguities of Technology, or Why Bobby Can't Spray', *Social Studies of Science*, 31, no. 4 (2001), 557–591.

8 Sheila Jasanoff, ed., *States of Knowledge: The Co-production of Science and Social Order* (London: Routledge, 2006).

9 Brian Balmer, 'The Social Dimension of Technology: The Control of Chemical and Biological Weapons', in *New Perspectives on Technology, Values and Ethics: Theoretical and Practical Discussions*, ed. Wenceslao J. Gonzalez (Dordrecht: Springer, 2015), 167–182.

10 The phrase refers here to Adele E. Clarke and Joan H. Fujimura, eds., *The Right Tools for the Job: At Work in Twentieth-Century Life Sciences* (Princeton, NJ: Princeton University Press, 1992) where they argue that scientific tools and techniques are not generated on the spot as fitting particular jobs but instead result from empirically observable social processes such as judging, choosing and testing.

11 We use the term 'constituencies' cautiously here, as it is clearly related to the concept of 'epistemic communities' that has been adopted within some areas of international relations scholarship. Hass defines epistemic communities as 'a network of professionals with recognized expertise and competence in a particular domain and an authoritative claim to policy-relevant knowledge within that domain or issue area'. Peter M. Haas, 'Introduction: Epistemic Communities

and International Policy Coordination'. *International Organization*, 46, no. 1 (1992), 1–35.

12 We will use the British Broadcasting Corporation (BBC) convention of alternating the politically sensitive names of the city: Londonderry and Derry.

13 A more detailed account, including the broader historical and political context, is discussed in Anna Feigenbaum, *Tear Gas: From the Battlefields of WWI to the Streets of Today* (London: Verso, 2017), 100–122. This monograph was published very shortly before our chapter was finalized.

14 Frederick Fraunfelder, 'Is CS Gas Dangerous?', *British Medical Journal*, 320, no. 7233 (2000), 458–459.

15 The National Archives (hereafter TNA), Kew, UK. CJ 3/66 Report of the Enquiry into the Medical and Toxicological Aspects of CS (Orthochlorobenzylene Malononitrile). Part I.

16 Cited in Douglas Black and John Gray, 'Sir Harold Percival Himsworth, K.C.B', *Biographical Memoirs of Fellows of the Royal Society*, 41 (1995), 200–218; see also Linda Bryder and Joan Austoker, eds., *Historical Perspectives on the Role of the MRC: Essays in the History of the Medical Research Council of the United Kingdom and Its Predecessor, the Medical Research Committee, 1913–1953* (Oxford: Oxford University Press, 1989).

17 Cited in John Gray and Christopher Booth, 'Harold Percival (Sir) Himsworth', *Munk's Roll*, Royal College of Physicians (1989–1993), 9, 238, available at http://munksroll.rcplondon.ac.uk/Biography/VolumeIX.

18 Renamed the Chemical Defence Establishment in 1970.

19 TNA, CJ 3/66 Report of the Enquiry into the Medical and Toxicological Aspects of CS (Orthochlorobenzylene Malononitrile). Part I.

20 Feigenbaum, *Tear Gas*, 100–122.

21 See Feigenbaum, *Tear Gas*, 120–122; for other examples of where the second Himsworth report has been invoked.

22 Brian Rappert, 'Health and Safety in Policing: Lessons from the Regulation of CS Sprays in the UK', *Social Science & Medicine*, 56 (2003), 1269–1278.

23 Steve Wright, '"Harmless Weapons" and Crowd Control', in *Shades of Deviance: A Primer on Crime, Deviance and Social Harm*, ed. Roland Atkinson (London: Routledge, 2014), 92.

24 Anna Feigenbaum, 'Riot Control Agents: The Case for Regulation', *The Sur File on Arms and Human Rights*, 12, no. 22 (2015), 101–113.

25 John Walker, *Britain and Disarmament: The UK and Nuclear, Biological and Chemical Weapons Arms Control and Programmes, 1956–1975* (Farnham: Ashgate, 2012), 46.

26 TNA CJ 3/66. Report on the Enquiry into the Medical and Toxicological Aspects of CS (Orthochlorobenzylidene Malonitrile). Part II. Enquiry into Toxicological Aspects of CS and Its Use for Civil Purposes. 1971.

27 Ibid.

28 Ibid.

29 For a discussion of STS and extrapolation see Brian Balmer, 'Shadow Values and the Politics of Extrapolation', *Science, Technology, & Human Values*, 42, no. 2 (2017), 311–314.

30 TNA CJ 3/66. Report of the Enquiry into the Medical and Toxicological Aspects of CS (Orthochlorobenzylene Malononitrile). Part I.

31 It is likely that this also followed on from the remit of the Committee to examine specifically medical effects of the use of CS.

32 Patrick Crogan, *Gameplay Mode: War, Simulation and Technoculture* (Minneapolis: University of Minnesota Press, 2011), xviii.

33 Sharon, Ghamari-Tabrizi, 'Simulating the Unthinkable: Gaming Future War in the 1950s and 1960s', *Social Studies of Science*, 30, no. 2 (2000), 163–223.

34 Crogan, *Gameplay Mode*, 59.

35 John Keegan, *A History of Warfare* (London: Pimlico, 2004).

36 Brian Rappert, 'Scenarios on the Future of Non-Lethal Weapons', *Contemporary Security Policy*, 22, no. 1 (2001), 50–74.

37 TNA, CJ 3/66 Note for the File. Himsworth Report Part II. n.d. Note of a meeting held at the Home Office 20 July 1971.

38 Rob Evans, *Gassed: Behind the Scenes at Porton Down* (London: House of Stratus, 2000), 276–282.

39 TNA DEFE 24/1915. Loose Minute. CR. A. W. Stephens. [Head of DS10], 1 October 1973.

40 TNA DEFE 24/1915. Letter David Young, Private Secretary to the Minister of State for Defence, to Michael Alexander, Assistant Private Secretary to the Secretary of State, Foreign and Commonwealth Office, 2 October 1973.

41 TNA DEFE 24/1915. CR and the Himsworth Report. IBC MacCleod, 10 October 1973.

42 In the area of security, see in particular: Kathleen M. Vogel, 'Framing Biosecurity: An Alternative to the Biotech Revolution Model?' *Science and Public Policy*, 3, no. 1 (2008), 45–54.

43 TNA CJ 3/66. Report on the Enquiry into the Medical and Toxicological Aspects of CS (Orthochlorobenzylidene Malonitrile). Part II. Enquiry into Toxicological Aspects of CS and Its Use for Civil Purposes. 1971. The committee had arrived at this conclusion by drawing on a test in which a standard 12.5 g cartridge of CS had been detonated in a 20 m^3 sealed room and the concentration of agent had been measured over time. The committee recommended a 5 g cartridge be developed.

44 TNA HO 287/1757. Armed Besieged Criminals. 16 October 1969.

45 TNA HO 287/1757. Use of Gas Grenades at 170 Overton Road, Kirkcaldy, on 19th October 1969, 21 October 1969.

46 TNA HO 287/1757. Letter VC Stewart, Scottish Home and Health Department to RA James, Home Office, 22 October 1969.

47 TNA HO 287/1757. ? to HMI. 7 November 1969.

48 Evans, *Gassed*, 276; Feigenbaum, *Tear Gas*, 120, points out that CS use set a precedent for introducing novel security measures, such as rubber and plastic bullets, with Northern Ireland being the test site. Additionally, the Himsworth Committee confined itself to uses in domestic unrest rather than battlefield situations. Yet, although it was the police who had deployed CS in the Battle of the Bogside, it also marked the first involvement of the military, brought in to support the police. The involvement of the British army had serious political overtones,

given Anglo-Irish history and the allegations from the Republican side that they were an occupying force further muddying the complex debate around CS use in Northern Ireland.

49 TNA CJ 3/128. Press Briefing. Draft Questions and Answers.

50 TNA HO 287/2215. Notes for Supplementaries.

51 Alex Spelling, 'Driven to Tears: The UK, CS Gas and the Geneva Protocol, 1969–75', *Diplomacy and Statecraft*, 27, no. 4 (2016), 701–725.

52 TNA CJ 3/66. Report on the Enquiry into the Medical and Toxicological Aspects of CS (Orthochlorobenzylidene Malonitrile). Part II. Enquiry into Toxicological Aspects of CS and Its Use for Civil Purposes, 1971. The committee argued that humans had evolved mechanisms for filtering smoke and other larger particles from the upper respiratory tract.

53 Spelling, 'Driven to Tears'.

54 From a biopolitical perspective, medical interventions designed to regulate populations are all examples of the state acting as a physician. See Thomas Lemke *Biopolitics: An Advanced Introduction* (New York: New York University Press, 2011).

Chapter 7

What Counts as a Chemical Weapon? The Category of Law Enforcement in the Chemical Weapons Convention

Michael Crowley

INTRODUCTION

Guided by the conceptual focus on distinctions and articulations of chemicals established in the introductory chapter of this volume, this chapter takes as its investigative focus the principle international regime through which state discourse regarding toxic chemicals is conducted, namely the Chemical Weapons Convention (CWC). Negotiated in the rare, fragile and short-lived spirit of optimism and cooperation between states in the immediate aftermath of the Cold War, the CWC's ambitious goal has been the identification, removal and destruction of the vast stockpiles of largely Cold War–era chemical weapons (CW) and associated production facilities; and the absolute outlawing of any subsequent development, acquisition and use of chemical weapons 'under any circumstances', while preserving the appropriate use of toxic chemicals for legitimate purposes. This chapter examines how effectively these latter goals have been realized by asking what counts as a chemical weapon in relation to interpretations of the category of law enforcement in the Convention.

In the first substantive section, the chapter discusses how the nature of toxic chemicals and the constraints upon their use have been articulated, and how distinctions between prohibited and non-prohibited purposes were initially drawn by the states negotiating the Convention. It investigates what is at stake within the ambiguous nature of the negotiated text in these critical areas and the resultant contested nature of subsequent state interpretation and implementation of the attendant obligations. It argues that these textual ambiguities and contested state discourses have significant 'real-world' consequences for both those potentially and actually targeted by weapons employing toxic chemicals throughout the world, with specific reference to the current (lack of effective) regulation of such weapons used in 'law enforcement'.

Within this context, the second section of this chapter analyses how the most common toxic chemicals employed in 'law enforcement', namely riot control agents (RCAs), have been articulated, and how demarcations between acceptable and prohibited purposes have been drawn under the Convention. The third section examines documented cases of misuse of such chemicals by police officers or prison guards, many resulting in serious injury and death. It suggests that such misuse may in part be legitimized or at least more easily ignored by states simply accepting 'law enforcement' as a legitimate purpose without examining the actions undertaken in its name. It argues that there is now the potential for the current situation – of ambiguities, contestations and consequent abuses – to worsen due to the ongoing development of a growing range of systems capable of inflicting significantly greater amounts of RCAs upon those targeted and to do so over wider areas or extended ranges.

Moreover, in addition to RCAs, certain countries have attempted to develop new weapons employing distinct toxic chemicals, notably so-called incapacitating chemical agents (ICAs) purportedly intended to act on the body's core biochemical and physiological systems to cause prolonged but non-permanent disability. Such activities are the focus of the fourth section of this chapter. By taking into account the contested nature of the discourse surrounding the legitimacy of ICA weapons as well as the contemporary state of research, development and use of ICAs, the chapter also highlights the potential threats these weapons pose for human rights, human security and the prohibition of chemical weapons. The chapter concludes with a summary of the current situation and brief examination of ways forward, suggesting the utility of a more contextualized, nuanced and holistic conceptualization of chemical control.

THE CHEMICAL WEAPONS CONVENTION

The CWC entered into force on 29 April 1997 and, on 23 March 2018, comprises 192 states parties that have ratified, acceded or succeeded to it[1] – the highest number of any comparable arms control or disarmament treaty. The CWC states parties represent about 98 per cent of the global population and landmass, as well as 98 per cent of the worldwide chemical industry.[2] The Treaty is of unlimited duration and was intended by the negotiating states to be far more ambitious and comprehensive in scope and application than any prior international agreement on chemical weapons.

Article I, which established the basic overarching obligations of states parties, explicitly prohibits the development, production, stockpiling and transfer 'directly or indirectly' of 'chemical weapons to anyone'.[3] It also expressly outlaws the use of chemical weapons 'under any circumstances' and requires their destruction within a specified time period.[4]

In the 2014 legal Commentary to the Convention, Krutzsch has argued that the chapeau phrase to Article I 'never under any circumstances' emphasized the comprehensive and totally binding nature of all the prohibitions set out under the subsequent paragraphs.[5] Krutzsch has contended that the wording of Article I has excluded any justification for the prohibited activities, and covered 'all intents and purposes for such activities, independent of the character of the armed conflict, whether, an international or non-international one, whether the parties involved had recognized themselves or whether or not it is a civil strife'.[6]

The comprehensive nature of the Convention was further established under Article II.2 which defined a 'toxic chemical' as: 'any chemical which through its chemical action on life processes, can cause death, temporary incapacitation or permanent harm to humans or animals'. Furthermore this definition explicitly includes 'all such chemicals, regardless of their origin or of their method of production, and regardless of whether they are produced in facilities, in munitions or elsewhere'.[7] Consequently, the Convention is not restricted to chemical warfare agents but covers toxic industrial chemicals, pharmaceutical chemicals or any other human-synthesized chemicals. Furthermore, it would also potentially cover naturally occurring chemicals, notably including toxic chemicals produced by living organisms (i.e. toxins).

The attempts by the negotiators to ensure a comprehensive scope of the Convention are also apparent in the Article II.1 definition of chemical weapons inter alia as 'toxic chemicals and their precursors, *except where intended for purposes not prohibited under this Convention, as long as the types and quantities are consistent with such purposes*' (emphasis added).[8] This highlighted text has become known as the General purpose criterion (GPC). In their 2014 Commentary to the Convention, Krutzsch and Trapp highlighted the implications of the open-ended nature of this provision:

> Under this concept, *all* toxic or precursor chemicals are regarded as chemical weapons *unless* they have been developed, produced, stockpiled or used for purposes *not* prohibited. The definition thus covers any toxic or precursor chemical if intended for chemical weapons purposes, irrespective of whether it has been listed in one of the Schedules [of restricted chemicals](Emphasis highlighted in original).[9]

Because the GPC establishes a prohibition based on intent rather than on a limited list of toxic chemical agents, it allows the Convention to accommodate and reflect developments in science and technology; consequently, as Meselson and Perry Robinson have highlighted, 'even toxic chemicals whose existence is not yet known are covered' by its provisions.[10] Meselson and Perry Robinson also noted that the GPC has a second important function;

it 'protects legitimate uses of all toxic chemicals and chemicals from which they can be made'.[11]

To determine whether the specific use of a toxic chemical would be in conformity with the CWC, the intention or purpose for its use needs to be established. The range of 'purposes not prohibited' under the CWC has been defined under Article II.9 as including, inter alia: '(d) Law enforcement including domestic riot control purposes'.[12] Consequently, any toxic chemicals held for purposes not provided for in Article II.9 are then considered to be chemical weapons and prohibited under the CWC. However, the use of toxic chemicals for 'purposes not prohibited' would be acceptable only 'as long as the types and quantities [of toxic chemicals] are consistent with such purposes'.[13]

Implications of Definitional Lacunae

While many of the terms utilized in the Convention are defined in the text, several important concepts were left undefined by the negotiators, notably including 'method of warfare', 'law enforcement' and 'domestic riot control purposes'. Certain commentators have argued that such ambiguity was intentional, accommodating the interests of governments such as the United States. For example, the March 1994 editorial of the *Chemical Weapons Convention Bulletin* noted: 'Some, by no means a majority, of the negotiating States wished to protect possible applications of disabling chemicals that would either go beyond, or might be criticized as going beyond, applications hitherto customary in the hands of domestic police forces'.[14]

Perhaps a degree of 'constructive ambiguity' was useful and even indispensable in developing a Convention text that all negotiating states could sign up to, given the opposing positions of some States on certain issues. Such textual in-exactitude is by no means unusual in international agreements. Where such ambiguity exists in international agreements, states have recourse to legal tools and guidelines to aid them in their interpretations of the text, most importantly the UN Vienna Convention on the Law of Treaties.[15] However, even utilizing such interpretive tools, the absence of definitional clarity and the consequent textual ambiguity that can result if left unresolved over a lengthy period of time can become dangerous, leading to differing interpretations of the CWC by states parties. This, in turn, could potentially lead to breaches of the Convention by some states parties and also to an erosion of the stability and key prohibitions of the regime as a whole. In short, there is a great deal at stake in the types of 'chemical bodies' that exist in these spaces of ambiguity.

Despite these ongoing issues, as of April 2018, no policy-making body of the Organisation for the Prohibition of Chemical Weapons (OPCW) has issued guidance on the interpretation of the meaning of 'law enforcement including domestic riot control purposes' and the nature and scope of activities covered under this term, nor established specific dedicated processes for doing so.[16] To date, it has been left to states parties to interpret the meaning of this term and the appropriate implementation of the Convention in this area. Indeed, there are 'real-world' consequences of this lack of definitional clarity for the regulation of law enforcement weapons, consequences of the forms of power and control that emerge from this uncertainty. Before turning to these, the next section will briefly indicate how the CWC definitional ambiguities have been further aggravated by the structural and cultural limitations of the international control regime.

Limitations of the Control Regime

The potentially deleterious consequences following from these ambiguities in the CWC text have been played out through, and further exacerbated by, the limitations in the structure, working practices and culture of the attendant control regime established to facilitate implementation of the Convention, namely OPCW. Indeed a number of commentators have highlighted the long-standing limited ability of the OPCW to address certain important or pressing issues, even if delays in action could potentially seriously weaken the effectiveness of the Convention.[17] Factors considered to have contributed to this situation have included the organization's culture of decision making by consensus and the consequent avoidance of difficult or controversial issues,[18] the wide disparity in resources and scientific and technical expertise available to state party delegations and national authorities, the limitations on the autonomy of the OPCW's Technical Secretariat including its ability to receive and act on open source information,[19] limited transparency and accountability of the OPCW to civil society and a reticence by the organization to receive information from and interact with relevant civil society organizations in a systematic manner.[20] The OPCW has on occasion found innovative ways to work around organizational obstacles – notably in its groundbreaking work facilitating destruction of declared CW facilities and stockpiles in Syria in the midst of the ongoing conflict – and has been attempting to address some of its systemic failings. However, a number of these factors have influenced how the organization has or has not reviewed and addressed the application of the Convention to ICA weapons, RCAs and related means of delivery.

APPLICATION OF THE CWC TO RIOT CONTROL AGENTS

RCAs, which include various forms of tear gas and pepper spray, are defined under Article II.7 of the Convention as: 'Any chemical not listed in a Schedule [of restricted chemicals], which can produce rapidly in humans sensory irritation or disabling physical effects which disappear within a short time following termination of exposure'.[21] In June 2013, the OPCW director general recognized that this definition of RCAs 'leaves some room for interpretation as to which chemicals can be considered as meeting the requirement specified'.[22] Consequently, in June 2014, the OPCW Technical Secretariat and Scientific Advisory Board developed a non-exhaustive and indicative list of seventeen toxic chemicals that corresponded to the Article II.7 definition and could therefore be considered as RCAs and subject to declaration under the CWC.[23]

In accordance with the definition under Article II.7, it is clear that RCAs are chemical agents that 'through chemical action on life processes, can cause death, temporary incapacitation or permanent harm to humans', and thus fall within the ambit of 'toxic chemicals' as defined by Article II, paragraph 2, of the Convention. Consequently, their use is permitted for a limited range of purposes defined under Article II, paragraph 9 and only 'as long as the types and quantities [of toxic chemicals] are consistent with such purposes'.

Although a range of toxic chemicals that fulfil the Article II.7 definition can be considered as 'riot control agents' under the Convention, this phrase belies the fact that RCAs have been used, in practice, for a variety of other potentially legitimate law enforcement applications beyond control or dispersal of rioting crowds, including for the incapacitation of violent individuals, personal defence, and in counter-terrorist operations.

Prohibition of Use of RCAs in Armed Conflict

In addition to the overarching constraints on the use of RCAs that flow from the regulation of toxic chemicals, the CWC under Article I.5 expressly prohibits the use of 'riot control agents as a method of warfare'.[24] Under international humanitarian law, the term 'method of warfare' is widely accepted as applying to both international and non-international armed conflicts.[25] Unfortunately, however, because the CWC does not explicitly define 'method of warfare' nor clearly demarcate the boundary between prohibited RCA use in warfare and permitted use in 'law enforcement', a recognized danger has been that states would develop varying interpretations of the Convention's restrictions in these areas. While the majority of CWC states parties currently appear to hold a comprehensive interpretation of Article I.5 prohibiting all military use of RCAs in armed conflict, one state party – the U.S. – maintains a long-held position, established in 1975 under Executive Order 11850, that RCAs can be legitimately used for a range of non-offensive actions, by military forces present in certain areas of armed conflict.[26] These include circumstances which could cross the line from 'law enforcement' into, or rapidly escalate to become situations of, armed conflict, notably where 'civilians are used to mask or screen attacks'; in the rescue of 'downed aircrews and passengers, and escaping prisoners' and 'outside the zone of immediate combat to protect convoys from civil disturbances, terrorists and paramilitary organizations'.[27] A further concern is that previous alleged use of RCAs in counter-insurgency operations, apparently in violation of Article I.5, notably by the U.S. military to clear caves in Afghanistan[28] and by Turkish armed forces against Kurdish fighters,[29] was never publicly challenged within the OPCW. As with other past incidents examined in this volume – notably in the chapters by Guillemin as well as Revill and Favero – the lack of challenge raises concerns that such practice may be repeated and potentially gain wider acceptance in the future.

Use of RCAs for Law Enforcement

As discussed previously, under Article II.9 (d), toxic chemicals are permitted for 'law enforcement including domestic riot control purposes'.[30] While this article clearly allows the use of RCAs by law enforcement personnel, the scope of permissible activities – beyond that of domestic riot control – is not defined and is contested. During the CWC negotiations, certain states outlined their interpretation of the meaning of 'law enforcement'. In May 1992, Ambassador Ledogar, head of the U.S. CWC negotiating team, described the U.S. position:

We understand the language 'law enforcement activities including domestic riot control' to mean that domestic riot control is a subset of law enforcement activities. We understand other law enforcement activities to include controlling rioting prisoners of war, rescuing hostages, counterterrorist operations, drug enforcement operations and non-combatant evacuation.[31]

However, at the conclusion of negotiations, no definition of the scope of 'law enforcement' was agreed among states, and the ambiguity remained. As the *Chemical Weapons Convention Bulletin* noted: '[Article II.9 (d)] fully protects the use of chemicals such as tear gas for domestic riot control. But what is "law enforcement"? Nowhere in the Convention is it defined. Whose law? What law? Enforcement where? By whom?'[32]

Because neither 'law enforcement' nor the types of activities covered by it have been defined under the CWC,[33] or subsequently by the OPCW policy-making organs, it is the obligation under international law of each individual state party to implement the relevant provisions of the Convention in good faith, taking into account 'any relevant rules of international law applicable in the relations between the parties'.[34] Of particular relevance are those rules and restrictions on the use of force by law enforcement officials – which can include police, security or military personnel – that arise from the wealth of existing international human rights law and related standards. Within this context, if RCAs (or indeed other toxic chemicals) are intentionally employed by law enforcement officials to carry out serious human rights violations in clear contravention of international laws and standards – such as through torture or ill-treatment – to violently suppress peaceful demonstrations, or to make lethal force more deadly, it is hard to see how such actions could be legitimate 'law enforcement' activities as permitted under Article II, paragraph 9 (d), of the CWC. Such an interpretation has been supported by certain recognized authorities in this area. Perry Robinson, for example, who served as an advisor to the UK National Authority for the CWC, believes that such intentional misuse of RCAs would be inconsistent with the CWC.

> RCAs are designed and intended to work by causing people to move away from a particular area, to depart from a riotous assembly, to stop being a mob . . . . The intention behind the use is important. To use RCAs to intentionally disable and injure people is not consistent with the CWC. Such use would only be consistent if you had a weird interpretation of the Treaty.[35]

However, it is far from clear that such an approach is accepted by CWC state parties. Instead CWC states parties treat 'law enforcement' activities as being de facto legitimate, without appropriately unpacking the term, examining the specific actions actually undertaken in its name or giving adequate acknowledgement of additional contexts (actors, intentions, power structures) associated with such chemical use and misuse.

Indeed, the failure by the states negotiating the CWC to define 'law enforcement' or of the OPCW subsequently to develop formal guidance in this area has resulted in the nature and scope of this permitted purpose becoming a shifting boundary term continuously negotiated informally through state practice. As will be discussed further, this has very serious consequences for those upon whom agents of the state use toxic chemicals in the name of 'law enforcement'; potentially allowing powerful and abusive actors to articulate (and mask) violent actions – and intentions – under the guise of this permitted purpose.

NATURE OF CONTEMPORARY MISUSE OF RCA BY LAW ENFORCEMENT OFFICIALS

When employed in accordance with manufacturers' instructions and in line with international human rights law and the CWC constraints, RCAs can provide an important alternative to other applications of force more likely to result in injury or death, such as firearms. Law enforcement officials throughout the world regularly use them to disperse violent crowds or to subdue dangerous individuals. Yet, they also are open to misuse. And such misuse appears to be widespread and serious.

A study conducted by the author of reports from the UN and regional human rights bodies and international non-governmental organizations (NGOs) has identified human rights violations committed by law enforcement officials utilizing RCAs in at least ninety-five countries or territories from 2009 to 2013.[36] The following sections provide illustrative examples of the types of human rights abuses committed by police, security or military officers and prison guards employing RCAs inappropriately under the guise of 'law enforcement'. Although a basic typology is employed to differentiate the nature of abuse, in reality, a specific incident may well fall under more than one category. This section therefore reveals the forms of RCAs misuse that have emerged from the spaces of the limitations in the CWC examined in the earlier section.

RCA Use in Torture and Ill-Treatment

The prohibition on torture and other cruel, inhuman or degrading treatment or punishment is absolute. It applies in all circumstances and, as part of international customary law, to all states. It is incorporated into numerous treaties including the Universal Declaration on Human Rights (UNDHR), the International Covenant on Civil and Political Rights (ICCPR) and the UN Convention against Torture. However, despite these obligations, the author's 2009–2013 survey recorded misuse of RCAs by law enforcement officials to facilitate or carry out such activities in at least eighteen countries, and this practice continues to be reported throughout the world. In the majority of recorded cases, police officers or prison guards employed hand-held chemical irritant sprays in a targeted fashion against individuals handcuffed, shackled or otherwise under their control often in prison cells or police stations. In addition, there have also been reported incidents, within and beyond such places of detention, where RCAs were employed against groups or crowds in a manner which could be deemed as 'collective punishment'. This includes the large-scale employment during 2013–2014 by Israeli security forces of tear gas, pepper spray and water cannon spraying malodorant 'skunk water' against Palestinian communities, in response to protests by some but by no means all community members against Jewish settlements in the West Bank.[37]

Use of RCAs for Suppression of Freedom of Expression, Association or Assembly

UN human rights monitoring bodies and international human rights NGOs have regularly reported the misuse of RCAs to intimidate or punish those involved in peaceful demonstrations in many countries, in violation of international law protecting freedom of expression, association and assembly, notably the UNDHR and ICCPR. The author's 2009–2013 survey recorded such instances in over seventy-five countries, and these practices continue and continue to receive widespread media attention. For example, the clearly excessive and often abusive use of tear gas and pepper spray by Turkish police and security forces attempting to suppress the anti-government 'Gezi Park' protests during the summer of 2013 drew international concern; much outrage focusing on the shooting and throwing of tear gas canisters into confined spaces such as homes, public shelters and even makeshift health clinics.[38]

RCAs Employed in Conjunction with Excessive Use of Force and Firearms

Although the 'inherent' right to life is enshrined in many international and regional human rights instruments, UN human rights monitoring bodies and international non-governmental human rights organizations have regularly documented the employment of RCAs as part of the indiscriminate, excessive or lethal use of force by law enforcement officials. The author's 2009–2013 survey documented such practices in at least eighty-three countries, often when used in conjunction with other 'less lethal' weapons or firearms. One recent example in February 2014 was the employment of RCAs against some 200 migrants, refugees and asylum seekers from sub-Saharan Africa who were attempting to swim to Ceuta, an autonomous city of Spain on the north coast of Africa, from its border with Morocco. Members of the Spanish Civil Guard opened fire with large rubber bullets, tear gas and blanks to stop their advance resulting in or contributing to the death of at least fourteen people. Survivors told Spanish NGOs that some of them were hit by rubber bullets when they were in the sea and that the tear gas fired by the Civil Guards made it difficult for them to see and breathe.[39] Although the details are contested, it appears that although the physiological effect of the tear gas may not have directly resulted in death, it may indirectly have contributed to the fatalities.

Serious Injuries or Death from Suffocation or Toxicity of RCAs

A recurring concern voiced by human rights monitors and the medical community has been the use of RCAs in excessive quantities in the open air or in confined spaces, including hospitals, prisons, homes and even automobiles, where the targeted individuals cannot disperse. In such situations, serious injury or death – particularly to the old, young and sick – can result from the toxic properties of the chemical agents or from asphyxiation. For example, in one notorious case of 18 August 2014, forty-five Egyptian prisoners were locked into a poorly ventilated cramped police van and transported to Abu Zaabal prison in Cairo. They were kept inside the van for hours in the prison yard without adequate water in the summer's heat, and when they shouted for help, a police officer fired a tear gas canister directly into the vehicle. When the van door was finally opened, thirty-seven of the prisoners were dead.[40]

In addition to contravening international human rights standards, the use of excessive amounts of RCAs in law enforcement operations appears to breach the CWC restriction on 'types and quantities' under Article II of the Convention.[41] According to Krutzsch and Trapp, in their legal Commentary to the CWC, RCAs can only legitimately be employed 'under circumstances that allow people to leave the place of exposure early enough before those effects become irreversible and or more severe than sensorily irritating or physically disabling'.[42] In certain situations, however, individuals are unable to leave the place of exposure in time, due to physical infirmity, age or the situation of the incident location (e.g. obstacles, RCA use in confined spaces or inside buildings). According to Krutzsch and Trapp: 'If the time span of exposure was so long that the sensory irritation or disabling effects no longer disappear spontaneously (or other toxic effects manifest themselves), the chemicals used changed legally from "RCA consistent with paragraph 9(d)" to "chemical weapons prohibited under Article I"'.[43] Consequently, Krutzsch and Trapp argue that states parties have to 'carefully apply rules to keep their use of RCAs in conformity with the "types and quantities" rule . . . . Taken together with other bodies of law (in particular human rights law), this requirement affects the acceptable tactics and modes of use of RCAs in law enforcement'.[44]

However, despite the ongoing widespread and serious misuse of RCAs by police and security forces, as of April 2018, no CWC state party has publicly raised any such cases as matters of concern formally within the OPCW, nor instigated any of the formal CWC consultation, clarification or investigation mechanisms available even though many such cases appear to breach either the 'law enforcement' provision (Article II.9 (d)) or the 'types and quantities' restriction (Article II.1 (a)) of the CWC. Furthermore, to date, no OPCW policy-making organ has addressed the nature and scope of 'law enforcement' under the CWC, clarified the range of permissible circumstances for RCA use and consequent constraints upon such use in line with CWC obligations and human rights law.

IMPLICATIONS ARISING FROM DEVELOPMENT OF WIDE-AREA RCA MEANS OF DELIVERY

If the OPCW does not take appropriate action to effectively regulate the existing use of RCAs in law enforcement, the situation could dramatically worsen as a result of ongoing development, marketing and potential subsequent deployment of a range of systems capable of delivering far greater amounts of RCA over wider areas or more extended distances than currently possible with standard law enforcement RCA dispersal mechanisms, such as hand-held sprays, grenades and single-projectile launchers. In addition to their potential misuse for collective ill treatment or punishment of crowds, such wide-area RCA delivery mechanisms could be employed as 'force multipliers' in conjunction with firearms, making lethal force deadlier on a large scale. Although nominally developed or acquired for law enforcement, they may also be incorporated into military arsenals, potentially resulting in subsequent use in armed conflict. Certain RCA delivery mechanisms, though designated as RCA munitions, could also be used to disperse other toxic chemicals such as classic chemical warefare agents. In addition, current weak trade controls in this area could result in acquisition of wide-area RCA delivery mechanisms and their misuse by non-state actors, including armed opposition forces, unregulated private military and security companies, criminal groups or terrorist organizations.

Investigations by the author and colleagues at the Omega Research Foundation and the Bradford Non-Lethal Weapons Research Project uncovered the development, production, possession or promotion by state or commercial entities in sixteen countries of 'wide-area' RCA means of delivery of potential concern since the coming into force of the CWC.[45] Companies in China, France, India, Israel, South Korea and the United States have developed a variety of multi-barrel launchers capable of firing large salvoes of RCA projectiles indiscriminately affecting substantial numbers of people.

Russian companies in particular reportedly have developed a disturbing range of RCA delivery mechanisms, including grenade rounds for automatic launchers, rocket-propelled grenades, mortar shells capable of reaching distances of more than 2.6 kilometres, munitions for use in mortars or artillery that could reach up to 6.8 kilometres, munition dispensers carried by helicopters and even cluster bombs packed with chemical irritant submunitions designed to be dropped from low-altitude helicopters or aircraft flying up to 12,000 metre above the target. Information on these RCA munitions was last made publicly available in 2009; it is not known whether production continues or stocks remain.[46]

A contemporary trend has been the development by companies or state research bodies in China, France, Israel, Spain, South Africa, the United Arab Emirates and the United States of unmanned ground vehicles or drones capable of carrying RCA-spraying devices or RCA projectile launchers. For example, the South African company Desert Wolf currently promotes the Skunk riot control drone. Capable of holding up to 4,000 pepper balls, this drone is equipped with four high-capacity paintball barrels that can cumulatively release '80 Pepper balls per second, stopping any crowd in its tracks'.[47]

Interest in acquiring Skunk drones apparently is not restricted to law enforcement bodies, with Desert Wolf claiming in June 2014 to have 'received an order for 25 units' for use by 'an international mining house', with additional potential customers including 'mines in South Africa . . . security companies in . . . and outside South Africa . . . and a number of other industrial customers'.[48]

Application of the CWC to Means of Delivery

Although the CWC under its definition of 'chemical weapon' included: '(b) munitions and devices specifically designed to cause death or other harm through the toxic properties of those toxic chemicals specified in subparagraph (a), which would be released as a result of the employment of such munitions and devices',[49] it did not list the types of acceptable or non-acceptable delivery systems for use with toxic chemicals in law enforcement scenarios. However, it did place an important potential constraint upon such means of delivery through the Article II.1 (a) 'types and quantities' restriction. In the 2014 Commentary on the Convention, Krutzsch has contended that 'the types and quantities restriction rule requires that, in order to be not prohibited, the technical characteristics and attributes of the means of delivery of RCAs must be consistent with the rules and conditions that apply to their use in law enforcement purposes under 9 (d) of Article II'.[50] Consequently, Krutzsch argues that the means of RCA delivery must 'be technically designed to enable law enforcement officers to use them safely for the permitted purposes. RCA delivery systems that are consistent with the permitted purpose of law enforcement including domestic riot control must allow for the limiting of exposure to a level acceptable for that purpose'.[51] Thus, accepting this legal analysis, it would follow that the employment of certain wide-area RCA delivery mechanisms may be justifiable in particular large-scale 'law enforcement' situations, provided they comply with the CWC provision on 'types and quantities' and are used in strict conformity with relevant human rights standards restricting the actions of law enforcement officials. In contrast, other, inherently military, forms of wide-area RCA delivery mechanisms that have been developed, such as artillery shells, aerial bombs, large-calibre mortar shells and cluster munitions, appear to be completely inappropriate for any form of legitimate 'law enforcement', having possible utility instead in large-scale human rights abuses or armed conflict. In this framing, these mechanisms become more akin to chemical weapons and therefore should be treated accordingly. Once again it is clear that textual ambiguities in the CWC concerning both 'types and quantities' and 'law enforcement', and the contested interpretations of their application have allowed a space for states and companies to develop, manufacture, promote and acquire a growing range of worrying RCA weapons under the guise of law enforcement, without any penalty or other sanction from the OPCW.

In recent years, the OPCW, notably its Scientific Advisory Board, began to examine the issues surrounding RCA munitions. In 2012, the board raised its concerns regarding 'reports of the commercial availability of munitions apparently designed to deliver large amounts of riot control agents over long distances'.[52] To date, however, few states parties have clarified their national positions in this area.

One notable exception has been Turkey. In September 2010, the Omega Research Foundation, the Bradford project and the South Africa–based Institute for Security Studies uncovered evidence of the promotion by the Turkish state–owned arms manufacturer Makina ve Kimya Endüstrisi Kurumu of its tear gas–filled 120 mm mortar round, the 2-chlorobenzalmalononitrile (CS) MKE MOD 251, at the Africa Aerospace and Defence exhibition held in Cape Town, South Africa, in September 2010. In February 2011, Uğur Doğan, the Turkish ambassador to the OPCW, confirmed that such activities were prohibited under the CWC and that this prohibition extended to other 'mortar ammunition containing tear gas or any other prohibited substance'. Turkey consequently reported its destruction of all remaining 120 mm CS MKE MOD 251 munitions, together with epoxy models and promotional materials, and took steps to halt the trade, promotion and brokering of such munitions.[53]

Turkey's robust actions in this area provided a powerful precedent for developing common understandings on this issue. However, as of April 2018, the OPCW has not established mechanisms to determine which forms of RCA delivery systems violate the Convention or to effectively regulate the trade and use of those that are consistent with the CWC. Consequently, this area remains one where the boundaries of prohibited and permissible law enforcement RCA delivery mechanisms are essentially open and fluid, being 'continuously negotiated' through state practice.

APPLICATION OF THE CWC TO INCAPACITATING CHEMICAL AGENT WEAPONS

Although the use of appropriate types and quantities of RCAs for law enforcement purposes is permitted under the CWC, certain countries have developed additional weapons employing other distinct toxic chemicals, notably so-called ICAs. Although there is no internationally agreed definition of ICAs under the CWC or elsewhere, they can be described as a disparate range of substances, including pharmaceutical chemicals, bio-regulators, and toxins that are intended to act on the body's core biochemical and physiological systems to cause prolonged but non-permanent disability. They include toxic chemicals that act on the central nervous system (CNS) to produce unconsciousness, sedation, hallucination, incoherence, disorientation or paralysis. With inappropriate doses, however, death can result. A variety of these agents are legitimately used for medical, veterinary or other peaceful purposes, but they could potentially be employed as weapons. To date there has been only one large-scale law enforcement use of such weapons (as detailed further).

Use by the Russian Federation

At around 5 a.m. on 26 October 2002, Russian *Spetsnaz* Special Forces, in their attempt to save 900 hostages held in a Moscow theatre by armed Chechen separatists, employed a secret ICA weapon. Following mass sedation of the occupants, the Special Forces stormed the theatre and shot all the separatists. Although the bulk of the hostages were freed, more than 120 of them were killed by the chemical agent,[54] and many more continue to suffer long-term health problems.[55] A trace analysis undertaken by researchers from the UK's Defence Science and Technology Laboratory (DSTL) of extracts of clothing and urine from survivors of the Moscow theatre siege concluded that the ICA weapon comprised a mixture of two anaesthetics, carfentanil and remifentanil.[56] To this day, the Russian authorities refuse to publicly disclose full details of the weapon they employed or provide any information regarding the nature and levels of such weapons they may have developed or stockpiled. Although individual CWC states sought further information from Russia on these matters through bilateral consultations, no CWC state party triggered the formal mechanisms for initiating an OPCW/CWC investigation of this incident.

Contested Utility and Continuing Research

Proponents of such weapons have long advocated their development and use in certain extreme law enforcement scenarios, such as the Moscow siege, where there is a need to incapacitate an individual or a group rapidly and completely without causing permanent disability or fatality. The proponents also have raised the possibility of using them as a tool in a variety of military operations, especially in locations where fighters and civilians are in close proximity or intermingled.

Scientific and medical professional associations, arms control organizations, international legal experts and human rights and humanitarian organizations, as well as a growing number of states, have criticized research and development of such weapons, contending that their use presents potentially grave dangers to health and wellbeing. The British Medical Association concluded that 'the agent whereby people could be incapacitated without risk of death in a tactical situation does not exist and is unlikely to in the foreseeable future. In such a situation, it is and will continue to be almost impossible to deliver the right agent to the right people in the right dose without exposing the wrong people, or delivering the wrong dose'.[57]

Disquiet about these weapons is further exacerbated by concern that rapid advances in relevant chemical and life sciences, particularly genomics, synthetic biology, medical pharmacology and neuroscience, will be harnessed to the development of such weapons. In a 2012 study, the Royal Society gave warning of 'active interest in performance degradation applications of neuroscience for both military and law enforcement purposes' and highlighted 'indications of interest among a number of States in the development and use of incapacitating chemical agents'.[58]

During 2013 and 2014, the author together with Professor Dando of Bradford University conducted a survey of research by state, academic and commercial entities potentially applicable to the study or development of ICA weapons.[59] The survey, published in October 2014, documented evidence of continued research in relevant fields by Russian scientists following the Moscow theatre siege. This included computer modelling of the application of 'calmatives' against groups of individuals in enclosed spaces[60] and exploration of the interaction of potential ICAs with human receptor sites.[61]

The survey also uncovered the Chinese development of ICA dart weapons employing an unknown anaesthetic agent for use against individuals. Such weapons were subsequently marketed by Chinese companies at international arms fairs held in China. In 2012, these weapons were reportedly held by the Chinese People's Liberation Army. China has provided no public information regarding its stockpiles of these weapons nor the specific purposes of their intended employment.[62]

Israeli security services employed an ICA weapon as an attempted assassination tool on at least one occasion, in 1997. However, there was insufficient publicly available information for Bradford University to determine whether any Israeli entity was subsequently undertaking research into weapons employing ICAs, or whether Israel – held stockpiles of such weapons. There was limited information available indicating that the Israel Institute for Biological Research may have been conducting work in potentially relevant dual-use fields, although the details of the specific research projects were not available.[63] The Bradford University survey once again illustrates how ambiguities in the CWC particularly with regard to the nature of 'law enforcement' and the 'types and quantities' of toxic chemicals that can be employed for such purposes have opened up spaces which certain states have exploited to undertake activities that appear to run counter to the aims and objectives of the Convention.

In addition, the Bradford University survey described how states including the Czech Republic, India, Iran, the United Kingdom and the United States had conducted dual-use research since 1997 that was potentially applicable to the study or development of such weapons.[64] For example, from 2005–2007, Czech scientists published papers describing their investigations over several years relating to a range of pharmaceutical chemicals including various opioids, ketamine, medetomidine and midazolam, specifically highlighting their potential utility as, in their words, 'pharmacological non-lethal weapons'. Research into such chemicals continued after 2007, but additional papers contained no explicit reference to their potential application as so-called pharmacological non-lethal weapons (NLWs).[65] The Czech Republic CWC National Authority subsequently investigated Czech research activities and in 2014 stated that 'there was no connection [between] the research [and] creation of any sort of weapons or devices which could be used for military or police purposes'. With regard to the Czech 'pharmacological non-lethal weapons' papers, the National Authority stated, 'Research programmes had justifiable medical goals, but their reporting in public media exceeded actual results of the research thus creating a false impression of possible development of some sort of chemical weapons'. Following their investigation, the Czech CWC National Authority subsequently 'organised several presentations for the Ministry of Health, Ministry of Defence (MoD) and the Ministry of the Interior with the aim to improve their knowledge of the Convention and the National Legislation for its implementation'.[66] As well as highlighting potentially dubious research activities, the Czech case also speaks to the difficulties that individual states, as well as the OPCW and the international community more broadly, face in attempting to police the boundaries of permissible and non-permissible activities in these areas. These difficulties of monitoring and facilitating effective national implementation of the CWC once again are being greatly exacerbated by the lack of clarity concerning correct application of the Convention, particularly with regard to the 'types and quantities' of chemical agents permissible and the extent of legitimate 'law enforcement' purposes.

Although the United Kingdom, the United States and some other countries that previously conducted research in these areas have now formally declared that they are not developing and do not possess ICA weapons, some states known to have developed and employed such weapons – notably the Russian Federation, China and Israel – remain silent.

Contested Interpretations of the CWC

As toxic chemicals, ICAs clearly come under the scope of the CWC; their use as weapons in armed conflict therefore is prohibited. Yet, there are differing interpretations among states, academics and legal scholars as to whether such weapons could be legitimately used for 'law enforcement' purposes under the Convention.

One line of treaty interpretation supported by Chayes and Meselson,[67] Krutzsch and Trapp,[68] the International Committee of the Red Cross (ICRC)[69] and a growing number of states parties including Australia, Canada, Germany, the United Kingdom and the United States hold that only RCAs can be used for law enforcement activities under the CWC. The only possible exceptions to this restriction recognized are those toxic chemicals used for judicially sanctioned execution, and only provided such chemicals are not on the Convention's Schedule 1 list of restricted chemicals. Consequently, the use of all other toxic chemicals, including ICAs,[70] as weapons for law enforcement would be prohibited. Krutzsch and Trapp, for example, argue that

> accepting, in principle, the use of certain toxic chemicals such as ICAs for law enforcement in exceptional situations, for instance during an anti-terrorist operation, would undermine the prohibition on Article I of the Convention and wrongly imply that there was a "choice" between applying binding rules of treaty interpretation and replacing them by explanations constructed as required by military interest.[71]

A second line of opposing interpretation, as advanced by Fidler,[72] and supported by Neill[73] and the International Law and Policy Institute,[74] contends that the CWC does not limit the range of toxic chemicals that can be used for law enforcement purposes to RCAs. Such interpretation may allow the use of ICA weapons for law enforcement in certain circumstances. However, Fidler notes that the 'use of a toxic chemical for law enforcement purposes is still subject to the CWC requirements that the types and quantities of chemicals . . . must be consistent with such permitted purposes (Article II.1 [a.])'.[75] According to Fidler, this restriction therefore 'requires scrutiny of the relationship between the chemical or biochemical agent and the law enforcement objective in question. The more difficult it is to control the effects of the use of a chemical or biochemical in a law enforcement operation, the more suspect such use becomes in terms of the agent being of a type or quantity consistent with a law enforcement purpose'.[76] Consequently, Fidler contends:

> For domestic law enforcement, use of incapacitating agents in contexts in which the government could control neither dosage nor the exposure environment would only be legitimate in extreme law enforcement situations . . . . For extraterritorial law enforcement activities undertaken by military forces and sanctioned by international law, States can at present only legitimately use riot control agents, not incapacitating agents.[77]

In recent years, there have been concerted attempts by a growing number of states, led by Australia and Switzerland, to encourage the OPCW to resolve the ambiguities concerning the development and use of these weapons. In December 2015, at the 20th Conference of States Parties (CSP-20), Australia issued a paper entitled *Aerosolisation of Central Nervous System-Acting*

Chemicals for Law Enforcement Purposes which was co-sponsored by twenty-one states.[78] The paper called on states parties to make known their national positions on the use of such chemicals in law enforcement and was intended to promote discussion within the OPCW, focusing on 'developing concrete recommendations for how to address [these] chemicals in a way that would significantly advance one of the OPCW's priorities – preventing the re-emergence of chemical weapons'.[79] The most recent iteration of this paper, issued at the 22nd Conference of States Parties (CSP-22) in November 2017, had by then acquired thirty-nine co-sponsoring states parties.[80] Furthermore, a growing number of individual states, including Australia, Canada, Germany, Switzerland, the United Kingdom and the United States, have now formally foresworn use of these chemical agents for law enforcement purposes and have continued to raise this issue in OPCW forums, with, for example, the U.S. National Statement to CSP-22 declaring that

> CNS-acting chemicals raise a new spectre of chemical weapons re-emergence. When it comes to these chemicals, the aerosolized use is not consistent with the law enforcement exception to the CWC as a purpose not prohibited by the Convention . . . . [T]he White House issued safety recommendations for first responders when handling and encountering fentanyl . . . . If our first responders are at risk when they encounter illicit fentanyl, how can our unsuspecting populations be safe when fentanyl is aerosolized and used as a law enforcement tool? The simple answer is that they cannot. Despite these dangers, countries continue to pursue these chemicals. If we do not seriously confront this issue here in the Hague, we would be turning a blind eye to the threat that CNS-acting chemical pose to the CWC – a threat that will increase, not decrease, over time.[81]

Regrettably, no policy-making organ of the OPCW has clarified whether these weapons can or cannot legitimately be employed for law enforcement purposes and, if so, under what circumstances and with what constraints. Once again CWC states parties are left to interpret the scope and nature of their obligations in this area, with the danger that a 'permissive' interpretation may evolve.

WHERE TO FROM HERE?

In the coming years, the OPCW is expected to oversee the destruction of the last declared elements of the largely Cold War–era state chemical weapons programmes. To remain relevant, the organization must embark on a process of reform and renewal to meet the new security challenges, to ensure chemical weapons remain beyond the pale and are never employed 'under any circumstances'. Critical to the success of this process will be whether and how the CWC states parties resolve certain controversial and contested issues that they have previously avoided, notably the development and use of weapons employing toxic chemicals for law enforcement purposes.

Although it can be argued that RCAs, when used in appropriate quantities and in strict accordance with human rights standards, can play a legitimate role in law enforcement (e.g. to disperse violent crowds), it is clear that they have been widely misused for human rights violations by law enforcement officials, and when used in excessive amounts have proved injurious or fatal to those targeted, particularly in confined spaces. While there is general consensus among CWC states parties that use of 'appropriate types and quantities' of RCAs for 'law enforcement' purposes would be in conformity with the CWC, to date the CWC states parties have been unable or unwilling to define these terms or to adequately address situations where RCAs have been used in clearly inappropriate quantities or in abusive activities, under the guise of law enforcement. In part, this inadequate response appears to be exacerbated by both the formal and tacit culture of the OPCW, eschewing controversy and conflict, being heavily reliant upon decision making by consensus. Furthermore, the very nature of the control regime established by states (as represented by their governments) to facilitate discourse on these issues primarily among themselves has sidelined and excluded civil society, most notably in this case those very people who have suffered from the inappropriate or abusive use of RCAs by law enforcement officials.

In addition to the accumulating evidence of widespread misuse of RCAs, concern within and beyond the OPCW has been growing about the attempts by certain states to develop completely distinct ICA weapons and to promote their utility, and argue for their permissibility under the CWC, in certain extreme law enforcement situations (notably anti-terrorist operations where large numbers of hostages are held captive). In contrast to the general acceptance by states of the appropriate use of riot control in law enforcement, no such consensus exists with regard to the employment of ICA weapons. Instead there is a concerted effort from scientific and medical associations, human rights organizations and a growing number of CWC states parties to clarify that such ICA weapons are prohibited under the CWC.

Despite the long-standing limitations of the OPCW, the author contends that, for the foreseeable future, it provides the most appropriate forum for the international community to address these tangled issues. Indeed, it can be argued that the OPCW, its member states and concerned civil society, in order to truly protect and nurture the Convention and enforce its core objective of ensuring the absolute prohibition of chemical weapons 'under any circumstances', are now obliged to seek a mechanism to address law enforcement weapons employing toxic chemicals. One potential mechanism would be for the OPCW to establish an open-ended working group (OEWG), to which all CWC states parties would be free to participate and which would be empowered to elicit scientific, technical and legal advice and information from relevant states, intergovernmental organizations and civil society organizations. The OEWG could be tasked with establishing clear guidelines designed to prevent research, development, production and employment of law enforcement toxic chemical weapons that would be in contravention of the Convention, and also establish clear guidelines as to the appropriate use of permissible law enforcement weapons (e.g. tear gas) and the constraints upon such use (e.g. a prohibition of the use of tear gas in confined spaces). In its deliberations, the OEWG should seek to employ a contextualized, nuanced and holistic approach to chemical control in this area, in particular seeking to examine and delineate the legitimate activities that should be considered to be 'law enforcement'. In so doing, the OEWG should consider existing obligations under relevant international law, notably international human rights law, and their bearing on the implementation of the CWC in this area. The OEWG would present its findings and recommendations to a relevant policy-making organ of the OPCW for adoption and subsequent implementation by all CWC states parties.

Failure to set clear boundaries in these areas risks the introduction and misuse of inappropriate weapons by police, security and military forces; the danger that such weapons will subsequently be employed for actions beyond law enforcement; and the threat of their proliferation to other state and non-state actors. In addition to the detrimental consequences for human rights and human security, such developments could undermine the stability of the Convention itself.

NOTES

1 An additional Signatory State (Israel) has signed the CWC, thus rendering political support to the objectives and principles of the Convention and committing itself to not undermining the Convention's objectives. Only four Non-Signatory States (Angola, DPRK, Egypt and South Sudan) have not taken relevant action on the Convention. The Convention text, details of states parties and all OPCW documents are available from OPCW website http://www.opcw.org.
2 OPCW, About the OPCW, OPCW website.
3 OPCW, Convention on the Prohibition of the Development, Production, Stockpiling and Use of Chemical Weapons and on their Destruction (Chemical Weapons Convention), 1993, Article I.1.a.
4 OPCW, CWC (1993), Article I.
5 W. Krutzsch, and R. Trapp, 'Article II: Definitions and Criteria', in *The Chemical Weapons Convention: A Commentary*, eds. W. Krutzsch, E. Myjer and R. Trapp (Oxford: Oxford University Press, 2014), 64.
6 Ibid.
7 OPCW, CWC (1993), Article II.2.

8 OPCW, CWC (1993), Article II.1.
9 Krutzsch and Trapp, 'Article II: Definitions and Criteria', 77.
10 M. Meselson and J. Perry Robinson, 'New Technologies and the Loophole', editorial, *Chemical Weapons Convention Bulletin*, 23 (Harvard Sussex Program, March 1994), 1.
11 Ibid.
12 OPCW, CWC (1993), Article II.9.d.
13 OPCW, CWC (1993), Article II.1.a.
14 Meselson and Perry Robinson, 'New Technologies and the Loophole', 2.
15 UN, Vienna Convention on the Law of Treaties, 23 May 1969, Vienna, 1155 UN Treaty Series (UNTS) 331.
16 Analysis was undertaken of all OPCW documents publicly available on the OPCW website (http://www.opcw.org) since the organization's inception up until 23 March 2018.
17 For example, see: Kelle, A. 'The CWC after Its First Review Conference: Is the Glass Half Full or Half Empty?', *Disarmament Diplomacy*, 71 (June–July 2003), 31–40; J. P. Perry Robinson, 'Non Lethal Warfare and the Chemical Weapons Convention', further HSP submission to the OPCW Open-Ended Working Group on Preparations for the Second CWC Review Conference, 24 October 2007.
18 W. Krutzsch and T. Dunworth, 'Article VIII: The Organization', in *The Chemical Weapons Convention: A Commentary*, eds. W. Krutzsch, E. Myjer and R. Trapp (Oxford: Oxford University Press, 2014), 258–260.
19 OPCW, Director General, Report of the [Ekeus] advisory panel on future priorities of the Organisation for the Prohibition of Chemical Weapons, S/951/2011, 25 July 2011, paragraph 48.
20 D. Feakes, 'Workshop Report: Pugwash Study Group on the Implementation of the Chemical and Biological Weapons Conventions CWC Implementation: Balancing Confidentiality and Transparency', (Noordwick, Netherlands: 15–17 May 1998); J. Tucker, 'Introduction', in *The Chemical Weapons Convention: Implementation challenges and solutions*, ed. J. Tucker (Washington, DC: Monterey Institute of International Studies, April 2001); M. Moodie, 'Issues for the First CWC Review Conference', in *The Chemical Weapons Convention*, ed. Tucker; A. Sands and J. Pate, 'CWC Compliance Issues', in *The Chemical Weapons Convention*, ed. Tucker; Kelle, 'The CWC after Its First Review Conference'.
21 OPCW, CWC, (1993), Article II.7.
22OPCW, Report of the twentieth session of the Scientific Advisory Board SAB-20/1, June 10–14, 2013, 14 June 2013, Annex 4, Director general's request to the Scientific Advisory Board to consider which riot control agents are subject to declaration under the Chemical Weapons Convention.
23 OPCW, Technical Secretariat, Office of Strategy and Policy, S/1177/2014, 1 May 2014, note by the Technical Secretariat, Declaration of riot control agents: advice from the Scientific Advisory Board. See also: OPCW, Report of the twentieth session of the Scientific Advisory Board SAB-20/1.
24 OPCW, CWC (1993), Article I.5.

25 The term 'method of warfare' refers to the way or ways in which a weapon is used. According to the Program on Humanitarian Policy and Conflict Research at Harvard University (HCPR): the term 'means of warfare' refers to 'weapons, weapon systems, or platforms employed for the purposes of attack', while the term 'methods of warfare' refers to 'attacks and other activities designed to adversely affect the enemy's military operations or military capacity, as distinct from the means of warfare used during military operations, such as weapons. In military terms, methods of warfare consist of the various general categories of operations, such as bombing, as well as the specific tactics used for attack, such as high altitude bombing'. Claude Bruderlein, *HPCR Manual on International Law Applicable to Air and Missile Warfare* (Bern: Harvard University 15 May 2009), 4–5.

26 President Gerald Ford, 'Executive Order 11850 – Renunciation of Certain Uses in War of Chemical Herbicides and Riot Control Agents, 40 FR 16187 CGR, 1971–1975 Comp., 980' April 8, 1975.

27 Ibid.

28 J. Fry, 'Contextualized Legal Reviews for the Methods and Means of Warfare: Cave Combat and International Humanitarian Law', *Columbia Journal of Transnational Law*, 44, no. 2 (February 2006), 453–519.

29 Sunshine Project, 'A Survey of Biological and Biochemical Weapons Related Research Activitiesin Turkey', *Sunshine Project Country Study*, no. 3, 8 December 2004, 15.

30 OPCW, CWC (1993), Article II.9.d.

31 See: written response by Ambassador Ledogar to the U.S. Senate Committee on Foreign Relations. [102d US Congress, 2nd Session, Senate, Committee on Foreign Relations, hearing, 1 May 1992, Chemical Weapons Ban Negotiations Issues, S.Hrg.102–719, 1992, 34–35].

32 Editorial, 'New Technologies and the Loophole in the Convention', *Chemical Weapons Convention Bulletin*, no. 23 (March 1994), Harvard Sussex Program, 1.

33 Law enforcement could be defined in ordinary parlance as 'the activity of making certain that the laws of an area are obeyed'. It is generally understood as referring to the exercise of police powers, especially with respect to the powers of arrest and detention, and may involve the lawful use of force. According to Chayes and Meselson: The term 'law enforcement' in Art. II (9) (d) means actions taken within the scope of a nation's 'jurisdiction to enforce' its national law, as that term is understood in international law. When such actions are taken in the context of law enforcement or riot control functions under the authority of the UN, they must be specifically authorized by that organization. No act is one of 'law enforcement' if it otherwise would be prohibited as a 'method of warfare' under Article II (9) (c). A. Chayes and M. Meselson, 'Proposed Guidelines on the Status of Riot Control Agents and Other Toxic Chemicals under the Chemical Weapons Convention', *Chemical Weapons Convention Bulletin*, 35 (March 1997), 15.

34 UN, Vienna Convention on the Law of Treaties, Article 31 (3) (a)–(c).

35 Professor Julian Perry Robinson in interview with author, 7 July 2009.

36 M. Crowley, *Chemical Control: Regulation of Incapacitating Chemical Agent Weapons, Riot Control Agents and Their Means of Delivery* (Basingstoke, Hampshire: Palgrave Macmillan, 2016), 50–70.
37 'Israel and Occupied Palestinian Territories: Trigger-Happy: Israel's Use of Excessive Force in the West Bank', Amnesty International, MDE 15/002/2014, 27 February 2014; 'Crowd Control Israel's Use of Crowd Control Weapons in the West Bank', *B'Tselem* (January 2013).
38 'Gezi Park Protests: Brutal Denial of the Right to Peaceful Assembly in Turkey', Amnesty International, EUR 44/022/2013, October 2013.
39 Amnesty International, News, 'Spain: Accountability Urged for "appalling" Migrant Deaths in Ceuta', 14 February 2014; Amnesty International, News, 'Spain: Ceuta Migrant Tragedy – Deplorable Disregard for Human Life', 6 February 2015.
40 P. Kingsley, 'How Did 37 Prisoners Come to Die at Cairo Prison Abu Zaabal?' *The Guardian*, 22 February 2014.
41 OPCW, CWC (1993), Article II.1.a.
42 Krutzsch and Trapp, 'Article II: Definitions and Criteria', 96.
43 Ibid.
44 Ibid.
45 For further discussion of these case studies and other RCA wide-area means of delivery of potential concern, including correspondence with relevant governments and companies, see: M. Crowley, 'Tear Gassing by Remote Control: The Development and Promotion of Remotely Operated Means of Delivering or Dispersing Riot Control Agents', (London: Remote Control Project, December 2015); Crowley, *Chemical Control*, 88–106; M. Crowley, 'Drawing the Line: Regulation of "Wide Area" Riot Control Agent Delivery Mechanisms under the Chemical Weapons Convention' (Bradford, UK: Omega Research Foundation/University of Bradford, April 2013).
46 M. Crowley, 'Drawing the Line', 25–38 (citing information from Non-Lethal Munitions Section, Volume 12 Ordnance and Munitions (English language version), Russia's Arms and Technologies, The XXI Century Encyclopedia, version 2006.1eng (Moscow, Russian Federation: Arms and Technologies Publishing House, 5 May 2009).
47 'Skunk Riot Control Copter', *Desert Wolf*, available at http://www.desert-wolf.com/dw/products/unmanned-aerial-systems/skunk-riot-control-copter.html.
48 L. Kelion, 'African Firm Is Selling Pepper-Spray Bullet Firing Drones', *BBC News*, 18 June 2014, available at http://www.bbc.co.uk/news/technology-27902634.
49 OPCW, CWC (1993), Article II.1.
50 W. Krutzsch, 'Article I General Obligations', in *The Chemical Weapons Convention: A Commentary*, eds. W. Krutzsch, E. Myjer and R. Trapp (Oxford: Oxford University Press, 2014), 71.
51 Ibid.
52 OPCW Conference of the States Parties, 'Report of the Scientific Advisory Board on Developments in Science and Technology for the Third Special Session of the Conference of the States Parties to Review the Operation of the Chemical Weapons Convention', RC-3/DG.1, 29 October 2012.
53 Crowley, *Chemical Control*, 102, 232–234.

54 J. Dunlop, *The 2002 Dubrovka and 2004 Beslan Hostage Crises, a Critique of Russian Counter-Terrorism* (Soviet and Post-Soviet Politics and Society 26), (Stuttgart: Verlag, 2006), 145–146.

55 M. Wheelis, 'Human Impact of Incapacitating Chemical Agents', in *Expert Meeting: Incapacitating Chemical Agents, Implications for International Law*, Montreux, Switzerland, 24–26 March 2010 (Geneva, Switzerland: ICRC, October 2010); D. Levin and V. Selivanov, 'Medical and Biological Issues of NLW Development and Application', Proceedings of the Fifth European Symposium on Non-Lethal Weapons, 11–13 May 2009 (Ettlingen, Germany: European Working Group on Non-Lethal Weapons, V23, 7); 'Four Years Later, Moscow Hostages Suffering', *CBS Evening News*, 21 October 2006.

56 J. Riches, R. Read, R. Black, N. Cooper and C. Timperley, 'Analysis of Clothing and Urine from Moscow Theatre Siege Casualties Reveals Carfentanil and Remifentanil Use', *Journal of Analytical Toxicology*, 36 (2012), 647–656.

57 British Medical Association, 'The Use of Drugs as Weapons: The Concerns and Responsibilities of Healthcare Professionals', *BMA Board of Science*, May 2007.

58 'Brain Waves Module 3: Neuroscience, Conflict and Security', *The Royal Society*, February 2012.

59 M. Crowley and M. Dando, 'Down the Slippery Slope? A Study of Contemporary Dual-Use Chemical and Life Science Research Potentially Applicable to Incapacitating Chemical Agent Weapons', Bradford Non-Lethal Weapons Research Project and Biochemical Security 2030 Project, October 2014, https://biochemsec2030dotorg.files.wordpress.com/2013/08/down-the-slippery-slopefinal-web.pdf.

60 See, for example: V. Klochikhin, A. Lushnikov, V. Zagaynov, A. Putilov, V. Selivanov and M. Zatekvakhin, 'Principles of Modelling of the Scenario of Calmative Application in a Building with Deterred Hostages', Proceedings of the Third European Symposium on Non-Lethal Weapons, Ettlingen, Germany, 10–12 May 2005, V17, Pfinztal: Fraunhofer ICT, 3–4.

61 For details of Russian research papers, see Crowley and Dando, 'Down the Slippery Slope?', 48–49.

62 Crowley and Dando, 'Down the Slippery Slope?', 14–20.

63 Ibid., 38–42.

64 Ibid., 20–37 and 50–72.

65 Ibid., 20–29.

66 Czech Republic, National Authority of the CWC, Reply to the University of Bradford, Re: Request for information concerning research potentially related to incapacitating chemical agents, July 14, 2014; for further discussion, see also: Crowley and Dando, 'Down the Slippery Slope?', 27–29.

67 A. Chayes and M. Meselson, 'Proposed Guidelines on the Status of Riot Control Agents and Other Toxic Chemicals under the Chemical Weapons Convention', *Chemical Weapons Convention Bulletin*, 35 (Harvard Sussex Program, March 1997), 13–18.

68 Krutzsch and Trapp, 'Article II: Definitions and Criteria', 73–104.

69 ICRC, 'Toxic Chemicals as Weapons for Law Enforcement: A Threat to Life and International Law?', September 2012.

70 Although there is no universally agreed definition of ICA weapons, for the sake of this discussion, they can be considered as weapons employing toxic chemicals whose purported intended use is to cause prolonged but non-permanent disability; they include centrally acting agents producing loss of consciousness, sedation, hallucination, incoherence, paralysis, disorientation or other such effects.

71 Krutzsch and Trapp, 'Article II: Definitions and Criteria', 102.

72 D. Fidler, 'Incapacitating Chemical and Biochemical Weapons and Law Enforcement under the Chemical Weapons Convention', in *Incapacitating Biochemical Weapons*, eds. A. Pearson, M. Chevrier, M. and Wheelis (Lanham, US: Lexington Books, 2007), 174.

73 D. Neill, 'Riot control and incapacitating chemical agents under the chemical weapons convention', *Defence R&D Canada Centre for Operational Research and Analysis*, June 2007.

74 International Law and Policy Institute, 'Chemical weapons and law enforcement under international law', ILPI Background paper, December 2016.

75 Fidler, 'Incapacitating Chemical and Biochemical Weapons', 174.

76 Ibid., 175.

77 Ibid., 185.

78 OPCW Conference of the States Parties, Aerosolisation of Central Nervous System-Acting Chemicals for Law Enforcement Purposes, C-20/NAT.2/Rev.1, 3 December 2015.

79 Ibid., paragraphs 10 and 12.

80 OPCW Conference of the States Parties, Twenty-Second Session, Aerosolisation of Central Nervous System-Acting Chemicals for Law Enforcement Purposes, C-22/NAT.5 (28 November 2017). The thirty-nine co-sponsoring states were: Albania, Argentina, Australia, Austria, Belgium, Brazil, Bulgaria, Canada, Chile, Colombia, Croatia, Cyprus, Czech Republic, Ecuador, Estonia, Finland, Germany, Greece, Ireland, Japan, Latvia, Lithuania, Luxembourg, Malta, New Zealand, Norway, Philippines, Poland, Portugal, Republic of Korea, Romania, Senegal, Slovenia, Spain, Switzerland, Turkey, United Kingdom, United States and Uruguay.

81 Statement by Andrea Hall, Senior Director for Weapons of Mass Destruction and Counter Proliferation, National Security Council, Delegation of the US to the 22nd Session of the Conference of States Parties, 27 November 2017.

Chapter 8

Tear Gas and Colonial Bodies in the British Interwar Period

Anna Feigenbaum

Bullet holes carved chunks out of the brickwork in the Jallianwala Bagh. Ricocheting off the walled enclosure of the garden in the city of Amritsar, at least 1,650 bullets were fired at protesters by British army officer Reginald Dyer's party – in only ten minutes. After their magazines emptied, Dyer ordered the soldiers to reload and fire again, directing them to target the densest part of the crowd. Confined by the architecture and trapped by armed forces blocking the garden's five exit routes, the 15,000 unarmed people gathered in the Bagh on 13 April 1919 had nowhere to run. They had gathered that day to resist the Rowlatt Act, an insidious piece of legislation that gave colonial authorities increased powers to detain, incarcerate and prosecute Indian subjects.[1]

By the end of the onslaught, the government's official inquiry identified 379 dead. Of this, 337 were men, 41 boys, and 1, six-weeks old. Non-governmental reports put the death toll higher – Indian Independence movement organizer Swami Shraddhanand wrote to Gandhi that 1,500 had been killed. Those who did not die of bullet wounds were trampled to death. The number of injuries was estimated at three times the number dead.[2] In the aftermath of the killings at Jallianwala Bagh, which came to be known as the Amritsar massacre, officials and news reporters struggled to make sense of the scene. Dyer tried to gain support for his actions from the many British who were committed to crushing the Indian Independence movement. But ultimately he was forced into retirement, his forces were retrained, and a firm stance was taken against the use of lethal force on non-violent protesters. By December 1919, the Associated Press was calling the Amritsar massacre the 'Darkest Stain on British Rule'. As recriminations over the bloody incident circulated, a question emerged on many local law enforcers' minds: What if the soldiers had used tear gas instead?[3]

Throughout the 1920s and 1930s, tear gas became intimately bound up with the British Empire's responses to calls for independence from its colonials. The dual function of tear gas as a physical (dispersing) and psychological (demoralizing) force meant that a certain amount of civil resistance to new rules and regulations could be easily contained. Furthermore, because these weapons were being legally used in the United States – and later in South Africa and parts of Europe – against peaceful or passively resistant demonstrators to make them 'look silly', authorities were no longer as threatened by new non-violent forms of collective action.[4] Tear gas was quickly becoming a go-to tool not only for suppressing the masses but also for calculatedly undermining acts of civil disobedience.[5]

In this chapter, I look at the tensions that surrounded the UK government's decision to deploy tear gas in its colonies. Drawing primarily from archival materials collected in the National Archive and India Office archives held at the British Library, this chapter takes a narrative approach to the telling of the conflicted motivations and concerns that shaped both policy decision around tear gas and the research and development of tear gas itself. Informed by actor network theory and postcolonial critique, the chapter explores correspondences between colonial officials, legal administrators and other influential figures, through a political economic lens. Weaving together these archival documents, I aim to show how tear gas was constructed and reconstructed in debates surrounding its use in the colonies during the interwar period. I argue that through the circulation of justificatory discourses between colonial administrators of the time, the framing of tear gas was transformed from that of a barbaric chemical weapon to a benevolent tool for policing civilian dissent. Drawing out the social as well as the political economic dimensions of this transformation, this chapter maps out how colonial bodies became sites for chemical intervention.

THE TEAR GAS DEBATES BEGIN

In February 1920, less than a year after the massacre in Punjab, the Punjabi government became one of the first colonial outposts to lobby for access to tear gas. News of the use of tear gas by the United States was spreading across the Atlantic as government and business travellers, along with international news articles, carried stories of the success of tear gas in suppressing 'riotious mobs'. After consultation with a self-proclaimed expert, the Punjab government discussed the possibility of utilizing lachrymatory gas to deal with mobs, and suggested a small supply of it might be maintained at certain centres in the province for use in emergencies.

Punjab officials brought their request to the Government of India, but in a confidential report circulated within the colonial offices, administrators worried that public opinion would strongly condemn the use of tear gas, which was still thought of as a device of war.[6] Great Britain had signed onto the 'Five Power Treaty' in Washington in February 1922, which contained Article V, banning gases in war. Upon signature, the India Office took the stance 'that the use of gas in war had been justly condemned by the general opinion of the civilised world, it was impossible for the Government of India to contemplate the employment of similar methods in times of peace for the dispersal of unlawful assemblies'.[7] In these early years of tear gas use, the U.S. public relations (PR) rhetoric was not enough to convince colonial authorities or the British Cabinet that the deployment of tear gas against civilians was worth the political risk.

For the British Empire, it was simply too soon after the Great War to consider poison gas beyond its horrifying wartime effects. Importantly here, the language of civilization was used to defend *against* the adoption of tear gas – a line of reasoning that was actively being countered in the United States by Amos Fries, chief of the U.S. Chemical Warfare Service (CWS), and his associates. While the dominant view in America claimed that chemical weapons (CW) were both the civilized option and a civilizing form of force, in Britain, this was still a minority opinion. But by 1926, when the Indian government revived the question of tear gas use, sentiment began to shift. Following the U.S. lead, international authorities came around to the position that war gases were inhumane, claiming that such views were outdated and did not account for the changes modern chemistry bestowed on the world. Like those advocating for gas use in the United States, colonial administrators began to leverage the loopholes in international agreements.

Back at the Washington Naval Conference meetings in 1921–1922, the Allied countries had come together to discuss, among other issues, the potential for a ban on chemical weapons in warfare. While they realized public opinion – and the experiences of their own soldiers – favoured a prohibition, they deemed this stance unrealistic on three grounds. First, many traditional weapons also used chemical processes in their explosive mechanisms. This would make it very difficult to differentiate between permitted and banned weapons. Likewise, they agreed that regulating the research and development of chemical weapons would be an impossible feat. Moreover, there would be no way to ensure other rogue and untrustworthy nations would follow the treaty. In order to be prepared for chemical attacks, it was important not to create a blanket prohibition. This line of reasoning carried echoes of the colonialist mindset. While the civilized Western nations could be trusted with chemical development, there was no telling what the 'savage' countries might do. This way of thinking about and representing the other served to justify the administration of violence by the British Empire over its colonies.[8]

With this easing on international regulations gaining credence, it became far simpler to argue for the right to use tear gas. In addition to its realpolitik attitude towards regulation, the Washington Treaty made no stipulation about the use of gas in times of peace for civil purposes. As self-proclaimed tear gas experts and marketing materials travelled from the United States into the British colonies, previous oppositions were increasingly countered with this new logic of legitimacy. 'So far as was known', a memo circulated to all local governments and the chief commissioner in the North-West Frontier Province of India read 'no case of panic among the crowd had ever occurred, nor was it apprehended that the release of gas would cause any more panic than might be produced by a discharge of musketry'. Setting the bar for entry low, the memo also set out experts' claims about the ease of tear gas use. In fact, 'the gas-gun could be handled by any ordinary intelligent man who had been given little instruction'.[9]

The case for tear gas adoption was furthered through appeals to the situation in Shanghai. Introduced to Shanghai by U.S. Marines, by the late 1920s, tear gas was held in the arsenals of the Shanghai police. There, according to H. O. DeGale, principal of the Police Training School in Phillaur, police found that 'the mere arrival of this highly trained body of men and their disciplined and precise action would, in the great majority of cases, be sufficient to cause the dispersal of the mob'.[10] These references to the situation in Shanghai also show how the circulation of news stories and gossip about tear gas travelled both between colonies and between the colonies and England. Anecdotal evidence of successful deployments were often coupled with reflections on the humane nature of tear gas and the promises it held for dealing with growing civil unrest. A letter published in the *Times* on 30 April 1930, signed only as from F.R.C.S. (from England), delivered news from South Africa, 'A recent disturbance among the coloured population at Durban in South Africa is reported to have been quieted by the use of tear-gas bombs'.[11] Such acts of circulation and distribution create new technological assemblages or networks. In a Latourian framework, they can be seen as acts of articulation, where the varied interests of actors are brought into connection with each other through acts of translation. Thierry Bardini offers this illustrative summary:

> Actors' identities and qualities are defined during negotiations between representatives of human and non-human actants . . . . The most important of these negotiations is 'translation', a multifaceted interaction in which actors (1) construct common definitions and meanings, (2) define representatives, and (3) co-opt each other in the pursuit of individual and collective objectives.[12]

Here, the needs of South Africa as a colonial empire are made similar to the needs of India.

In another circulated clipping on the benefits of tear gas, an ophthalmic surgeon reported that from his clinical experience, while many veterans claimed that gas exposure impaired their vision, the surgeon found that they were just older and needed glasses. It was only in those cases where drops of irritant liquid had gotten into the eye that the doctor noticed a direct cause of impairment. The surgeon concluded that the advantages of tear gas were 'obvious enough'. This 'more humane and scientific procedure' should surely replace 'older pre-war methods'.[13]

Over time, this PR strategy of enlisting medical professionals in the campaign to legitimate tear gas deployment grew commonplace. What better way to turn a poison gas into a scientifically advantageous toxin than with the recommendation of a doctor? It is quite difficult to imagine a medical professional advocating the health benefits of bullets or batons. Yet its status as a chemical meant that tear gas could be measured along different criteria than other police weaponry. It existed not only in the realm of tool construction but also in a world of modern chemistry – a science of the civilized world.

Unlike earlier efforts that tried to distance tear gas from the poisonous chemistry of the Great War, these medical men argued that gases, like drugs, existed on a continuum of harm and risk. Through experiment and examination, by monitoring trials and side effects, they could yield insight into the relative safety of tear gas. As these men of science lent their credibility to tear gas, the wood and metal weapons of the past could not compete. While this chemical fix-me-all marketing won over enough public opinion in the United States to outplay the opposition, the British Cabinet was not as convinced that the colonies were ready for tear gas. Over a decade after the United States and other nations first introduced war gases for peacetime use, why did the United Kingdom continue to hesitate? The mounting protests in the colonies were forcing debate between the War Office, Colonial Office, Army Council and British Embassies. To make matters more complicated, staff changeovers and political elections meant a rotating set of people were involved in decision making. As memos and letters bounced back and forth between departments, the debate over tear gas appeared to move back and forth in time. There was no centralized news service or collation of opinions. Rather, there was a string of documents that moved in messy transactions between officials, who then wrote more opinions that joined the pile.

This created a winding, start-and-stop style conversation that left a long list of questions unanswered. How should international policy be interpreted? Was tear gas a sign of inhumane treatment, or the greatest invention to ever grace civilian control, with its promise of 'no permanent injuries'?[14] Did the colonies have the skills and resources needed to expand its policing arsenals, or would rolling out tear gas reveal further incompetence in law enforcement? Rather than dismiss tear gas requests outright, the British Cabinet drew up a list of suggestions to take forward. With reference to the poison gases of the Great War, cabinet members argued that the public would require a great deal of education on the nature and effects of tear gas. In particular, it recommended, 'a different nomenclature should be adopted for the two kinds of gas'.[15] Giving 'tear gas' a new name would help differentiate chemicals used on civilians from those used on the battlefield. They further recommended that British delegates bring this nagging question about the exemption of tear gas use on civilians to the next international policy conferences on chemical warefare in Geneva and Washington. With a general election around the corner, the cabinet felt the return to a Conservative government might indeed sway feelings on tear gas in a more positive direction.[16]

On the backdrop of an economic depression and unemployment crisis back home in the United Kingdom and growing movements for independence in the colonies, the 1929 election brought a hung parliament, with Labour coming into power. But that power would not last long. With both the Labour and Liberal parties divided over how to respond to the depression, Prime Minister MacDonald formed a coalitional government. While this left him at the helm until 1935, Labour lost four out of every five seats, giving the Conservatives a large majority and delaying commitments to work towards independence of India, underway in the Socialist-Labour government of the late 1920s.

PASSIVE RESISTANCE AND THE 'WOMEN PROBLEM'

Amid this political climate, and with the rise of Satyagraha, it was not hard to convince colonial authorities in India on the benefits of tear gas. Satyagraha incorporated non-violent tactics including non-cooperation with tax systems, strikes from work, sit-ins, peaceful barricades and, in the words of the authorities, 'endeavours to prevent movement of Government officials or forces by prostrating themselves across roads etc.'[17] At the same time, Satyagraha also involved fasting and prayer as a means of purifying and connecting the individual and collective. Translated into English in a few different ways, 'Satya' as 'truth' refers to practice of seeking truth by trying to develop an understanding of a situation or conflict from all points of view. This 'truth' was not something that one could capture or own, but rather a pursuit or an orientation to understanding conflict.[18] Satyagraha was rapidly growing and developing in India, while colonial authority was slipping away.

During the twenty-four-day, 240-mile – long Salt March from Sabarmati Ashram to Dandi, beginning on 12 March 1930, protesters practised Satyagraha. An act of economic refusal and rejection of the colonial administration, the Gandhi-led salt march drew international media coverage from foreign journalists and garnered the support of thousands of Indians along the way. These Satyagraha tactics left the police in a diplomatic bind. Firing on protesters as they did in Amritsar would mean facing heightening tensions and damning press coverage, while doing nothing would risk the loss of control of a population increasingly engaged in both social and economic acts of non-cooperation. In a briefing memo to Lord Passfield in April 1930, the socialist appointee to the role of secretary for the Colonies was advised, 'Fire-arms can hardly be used in such circumstances', which led to 'considerable difficulty' for the police.[19] Known to his friends as the prolific Sidney Webb, Lord Passfield was a member of the Fabian Society and a founder of the London School of Economics and New Statesman. The Fabian Society was committed to non-violence and believed in evolutionary rather than revolutionary socialism. They believed in building a socialist democracy through local governmental organizing, rather than through any use of force. Historian Mary Davis argues that Lord Passfield exemplified the 'white man's burden' approach to colonial rule that saw it as Britain's duty to keep order in the colonies.[20] From this perspective, there was no better tool to do this with than tear gas.

In a May 1930 meeting to again consider India's desire to deploy tear gas, the British government stuck to its prohibition on tear gas for the colonies. The fear of 'moral disapprobation' was still enough to persuade the British government that it was not yet time to adopt tear gases for policing. While the government conceded that there was evidence for its effectiveness, the repercussions that might follow from its use outweighed those benefits – at least for the time being.[21]

Meanwhile, requests for tear gas also began to come in from Nigeria. As in India, this unrest brought problems for the British, who were reluctant to use violent and lethal force against women. As a strategy for administering colonial order in parts of Nigeria, the British had put men in charge as land chiefs, creating an abusive patriarchy that enforced property and tax laws. Women, angry at these acts of gendered and economic exploitation, joined together in protest. A memo updating colonial authorities of the situation read, 'Recently in Nigeria a hostile mob was composed largely of women, and the local troops showed the greatest dislike in firing on the crowd when that course became inevitable'.[22] This incident was part of what become known as the Women's War in Nigeria. This rising role of women in protest and the frustrating practice of 'Satyagraha' or passive resistance in India (and spreading farther abroad) posed a logistical and PR nightmare for colonial authorities. They needed an option that could quell resistance on the streets, while placating concern abroad. Tear gas offered a third way out – it could change how governments looked, without any need for them to change the way things actually were.

Yet two arguments against its use remained as hurdles. First, there was the 'moral difficulty' that came from the association of tear gas with other war gases like the deforming and often deadly mustard gas. As Britain denounced Germany's use of poison gas in World War I, and had signed onto the Washington Treaty in 1922 and the Geneva Gas Protocol in 1925, the Lord Privy Seal felt a 'position that lachrymatory gas may properly be used in the case of civil disturbances but not in war is impossible to maintain'.[23] However, the War Office felt this could be mediated by 'a declared intention to use tear-gas and adequate warning given to the opponents'[24] – a protocol that would in fact become adopted by the United Kingdom, emblemized in the Empire's large lettered banners reading: 'Disperse or we fire Tear Smoke'.

SUPPLYING THE NEW DEMAND

At this point, the UK's reluctance to open the tear gas market was not only political but also financial. With an incredibly limited remit to manufacture tear gas at home, Britain was lagging behind in the innovation and stockpiling of riot control supplies. While the French and Germans led the development of lachrymatory agents during World War I, after the war, it was the United States that was seen as the primary supplier of the world's tear gas. The entangled nature of U.S. military diplomacy and entrepreneurial manufacturing dominated the international market, while domestic production continued on smaller scales around Europe. While not yet in the manufacturing game, the British Empire had highly developed chemical industries and could 'produce effective gases easily and quickly'.[25]

However, for the United Kingdom to enter the tear gas market, it would need to capitalize on the chemical agents' economic efficacy. Touted in a briefing document for the British Cabinet as particularly effective for its ability to reach 'every member of a crowd, whereas the lethal weapon is individual in effect', tear gas was positioned as a weapon that could reduce police spending. Drawing on promotional material for tear gas, officials argued that tear gas could be used in an earlier stage of a riot than as a lethal weapon, which would lessen the burden of responsibility placed on officers. The possibility for developing a new exports and trade industry appealed to the business interests of government officials.

By the start of the 1930s, officials were coming around to the use of tear gas in the colonies. J. E. W. Hood noted to his colleagues in the colonial administration on 12 August 1930 that the latest War Office memo made a very clear and convincing case for the use of tear gas. 'I must say that I should prefer to use tear gas on a mob rather than shoot at them', Hood wrote, 'Tear gas does no harm but makes the mob look silly'. Hood noted that tear gas was used in the United States and South Africa without 'a howl of protest', reasoning, that 'if you may use gas on a Hottentot why not on an

Ibo'. Hood's remarks reflected the dominant colonial perceptions of the time, what we would now call an Orientalist view that saw native populations as inferior to white settlers, a population to be managed by the civilizing force of the West.[26]

In 1935, the new secretary of state for the colonies, Malcolm MacDonald, was given the task of sorting out an overarching policy on tear gas. To help brief him for this task, MacDonald was supplied an extensive memo. It supplied a laundry list of successful tear gas deployments, including Germany (used effectively against strikers in Hamburg in 1933), Austria (to dispel communist rioters in 1929), Italy (police were now armed with tear gas bombs as part of their standard equipment) and France (where permission for use was already granted). The memo also featured South Africa, attaching a clip from the 14 November 1929 edition of the *Evening Standard*:

> Hundreds of police officers today raided the native barracks in Durban, and searched 6,000 natives for poll-tax receipts. The police were equipped with tear-gas bombs, gas masks, machine guns rifles and bayonets. One gas bomb was used, with immediate effect, on a group of defiant natives. Six hundred natives were arrested, either for non-payment of the poll-tax or truculent resistance to the police.[27]

Under MacDonald's tenure, South Africa and the U.S. aggressive approach to tear gassing protesters would become a benchmark for success, rather than a cause for concern.

The aforementioned examples of the early 1930s cautious rhetoric that first suggested tear gas only be used in extreme cases, when firearms were about to be drawn, no longer applied. Instead, colonial officers were advised to fire gas early, to use it in large amounts and to protect the police through keeping distance and donning gas masks. Tear gas proved to be effective as a propaganda tool. Its great 'moral effect' could put down striking workers and dissenting minorities alike. Consequently, by the end of the 1930s, tear gas became the humane weapon of choice across the British Empire. Yet, while Britain underwent this ideological shift on tear gas over the course of the 1930s, the Empire had not yet prepared to produce British-made tear gas products for policing its dominion. The inspector general of police in Singapore wrote to the British Crown Agents, 'I am in possession of particulars of American made equipment (gas guns, truncheons and grenades) but am anxious to explore the possibilities of the Home market'. What good was the right to use tear gas without a manufacturing market to supply the demand?[28]

With little assistance from the War Office to secure supplies of tear gas, stocking India's arsenals fell into the hands of prominent businessmen. During the 1930s, Messrs. Govan Brothers and its managing director Grant Govan were given permission to open a depot in Bombay for storing tear gas to supply to the provincial governments.[29] An avid pilot and entrepreneur with an eye towards colonial industry, in 1922, Govan had incorporated Govan Brothers, Ltd. setting up *Delhi Flour Mills* and *Gwalior Sugar*. Govan also established the domestic Indian National Airlines and later a travel agency. Serving as the founding president of the Board of Control for Cricket in India, Govan's links to the colonial elite through travel, sport and industry, well positioned Govan Brothers, Ltd. at the intersection of policing, colonial authority and industry that, as in the United States, shaped the early years of tear gas as it moved into a commercial market for domestic policing.[30]

With the start of World War II, managing riot control imports and training became even harder. The colonial authorities tried to set up a system wherein provinces would send information on the amount and type of supplies needed to Govan Brothers, Ltd., which would then try to fill these requests, importing from the United States. But this was not an adequate solution in the long term. Civil disobedience in India was mounting once again. While provinces in India now had a number of trained gas squads and enough tear gas for immediate use, a larger problem remained. Tear gas, like other chemical substances, has a limited shelf life.

Authorities found that the ammunition deteriorated fairly rapidly and could not be relied on for a period of more than three years. This meant that it was not enough just to supply tear gas once, but that stocks had to come in regularly. As supplies were coming from so far abroad, India relied on stock reserves like the ones set up by Govan Brothers, Ltd., but the expiration of tear gas meant that it was difficult to rely on such a system.[31] The deterioration of equipment also meant that expired stock either needed to be used as practice ammunition, or it would need to be destroyed. Both of these requirements put additional pressure on the British to establish an infrastructure for riot control manufacturing and distribution.

THE LETHAL NON-LETHAL WEAPON

Hoping to manufacture British-made tear gas equipment for the colonies, scientists sent the War Office's 92 grenade out for trials. While most police forces in India were supplied with U.S.-made tear gases, Peshawar Central Jail stocked the British 92 grenades. In May 1947, on a hot sultry day with still winds, guards deployed these grenades in the jail's crowded prison yard. The jail filled up after conflicts over the treatment of Muslims in Peshawar came to a head. Acts of dissent ranged from civil disobedience including marches, strikes and sabotaged transport to more violent actions of stabbings, arson and attacks on police. By the end of April 1947, the police had detained nearly 5,000 Muslim activists. Jails overflowed forcing authorities to create temporary accommodation for the record number of political prisoners.[32] On 20 May 1947, the enclosed courtyard of the Peshawar jail contained 1,100 men. Prisoners crowded together by the railings. Hostility with the Peshawar police festered in the heat.

Stones and shouts of abuse filled the stagnant air of the packed prison courtyard. Then the police fired tear gas grenades. The first round filled the entrapped space with smoke, making it impossible for the prisoners to stay along the railings. They moved, choking and stumbling to another spot. Then, more grenades were thrown forcing prisoners to flee back into their cells. During the altercation, prison guards threw thirty-eight British 92 series grenades. Without wind and in a confined architecture, the toxic smoke of the tear gas persisted for four to five hours, with lingering effects causing discomfort for sixteen hours after the attacks. Shards of the grenades were judged to have exploded in a twenty-five to thirty yards radius. Of the 1,100 prisoners inside the courtyard, nearly all were affected by the gas. Many were overcome by the fumes and had to be carried out into non-affected areas.[33]

In a medical report on the incident, Peshawar's civil staff surgeon outlined three sets of clinical effects from the gas. First, there were those affected by the vapour alone. They suffered from lachrymation, conjunctivitis, cough, giddiness, nausea, vomiting, mental depression and unconsciousness of short duration. The surgeon noted that the effect 'varied with the dosage of tear gas received'.[34] Second, there were a total of twenty-eight cases of wounds, 'either multiple "peppering" or cleanly incised or punctured wounds of little depth'. These wounds were likely caused by shards of the grenades as they exploded and splintered to release the gas. The surgeon noted that the largest deep wound ran 2.5 inches long across the back of a patient's calf. In addition to this deep puncture wound, there were six cases of 'retained foreign bodies' in which thin metallic fragments became lodged into people's flesh.[35] The third set of injuries described thirteen cases of skin lesions with first and second degree burns. Then came the description of one man's death.

Arriving to hospital after the tear gassing, the patient 'had extensive second degree burns on the lower limbs, especially the right leg below the knee: and first degree burns on the trunk and upper limbs'. These burns included 'small peppered wounds' that covered about 30 per cent of the total body surface. While he was treated for these wounds, at the time the man showed no signs of severe distress. Two days later, at around 7:30 pm on the 23 of May, the prisoner collapsed. Vomiting blood (haematemesis) and passing tarry black stools (maleana), the man died four hours after being taken to the doctor. But before the surgeon could perform a post-mortem, the body was 'seized by unauthorized person, so the cause of death was not precisely determined'.[36]

Reviewing the police and surgeon's reports, the War Office determined that contrary to reports that tear gas could cause no lasting harm, the No. 92 grenade was indeed 'dangerous, possibly lethal, in its effect'.[37] Supplies of series 92 grenades were halted and a quest for alternatives began.

FROM COLONIAL BODIES TO CHEMICAL BODIES

This colonial history of tear gas tells us many things about how technologies are always entangled with other humans and non-humans. From the stagnation of wind in an Indian prison to the financial profits of Rhodesian mines, to understand the history of tear gas requires a tracing of interests and articulations. It is in trying to create narratives of acts of translation – those moments that Latour argues bring actors and actants together in a network around a common goal – that a history of tear gas emerges.[38] As Britain's post-war position shifted from trying to distance itself from the horrors of wartime gas to needing a less lethal solution to quell rising independence movements, its political alignments, justificatory discourses and eventually policies changed. For the United Kingdom to attempt to maintain the status quo of administrative rule, they needed to break the collective will of the people – to poison it, dissipate it and, most importantly, do so without much bloodshed. It is in this colonial moment that this chemical weapon, already embraced by the United States, South Africa and much of Europe, became the supposedly humane and civilized solution to social control.

In shifting from a condemned weapon of warfare to an embraced technology for repressing dissent, tear gas rendered colonial bodies as chemical bodies – bodies that could be gassed, that could be governed through the air. This re-imagining shifted focus from the baton or bullet used to break the individual to a means for breaking relations between individuals. In poisoning the sky – or threatening to – colonial administrators turned collective resistance into a site of vulnerability. The sit-down protests, the human lines passively refusing to move, needed to be dispersed and demoralized. By making the air uninhabitable, the street toxic law enforcement was able to make the very act of assembling bodies together impossible. In constructing and administering this new logic of public order policing, tear gas gave way to the string of 'less lethal' developments that followed, from the flash bang grenade to the rubber bullet. Able to appear benevolent, while maintaining control and invoking a prohibitive terror, a warning to future protesters, the rise of tear gas marked a new era of atmospheric policing that would come to spread across the world.

NOTES

1 N. Collett, *The Butcher of Amritsar: General Reginald Dyer* (London: A&C Black, 2006); N. Lloyd, *The Amritsar Massacre: The Untold Story of One Fateful Day* (London: I.B. Tauris, 2011).

2 Collett, *The Butcher of Amritsar*.

3 Dyer admitted in testimony that he had premeditated intention of firing on the crowd: 'I think it quite possible that I could have dispersed the crowd without firing. . . . But they would have come back again and laughed, and I would have made, what I consider, a fool of myself. . . . Therefore I considered it my duty to fire on them and to fire well'. Cited in Collett, *The Butcher of Amritsar*, 333–334.

4 British National Archive, Nigeria, CO 323/1113/12.

5 The term 'tear gas' was predominantly used in political correspondences of this time. Occasionally the term 'lachrymatory agent' is used, referring to those chemical compounds that were marked by making people tear or cry. There were a variety of early compositions of tear gas, with chloroacetophenone (CN) and later 2-chlorobenzalmalononitrile (CS) becoming the dominant form. Today, a variety of different chemical compounds continue to be referred to under the broader umbrella of tear gas.

6 British Library, India Office Archives, Use of Tear Gas in India IOR/L/PJ/12/414 [1927–1933].

7 Ibid.

8 Edward W. Said, *Orientalism: Western Conceptions of the Orient* [1978] (Harmondsworth, England: Penguin, 1995).

9 British Library, India Office Archives, Use of Tear Gas in India IOR/L/PJ/12/414 [1927–1933].

10 Ibid.

11 Ibid.

12 Thierry Bardini, 'Bridging the Gulfs: From Hypertext to Cyberspace', *Journal of Computer-Mediated Communication*, 3, no. 2 (September 1997), available at http://jcmc.indiana.edu/vol3/issue2/bardini.html#Agents.

13 British Library, India Office Archives, Use of Tear Gas in India IOR/L/PJ/12/414 [1927–1933].

14 Ibid.

15 Ibid.

16 Ibid.

17 British National Archive, Nigeria, CO 323/1113/12.

18 Daniel M. Mayton II, 'Gandhi as Peacebuilder: The Social Psychology of Satyagraha', in *Peace, Conflict, and Violence: Peace Psychology for the 21st Century*, eds. D. J. Christie, R. V. Wagner and D. D. N. Winter (Upper Saddle River, NJ: Prentice Hall/Pearson Education, 2001), 307–313, available at http://u.osu.edu/christie/files/2014/10/Chapter-25-Gandhi-as-Peacebuilder-Mayton-2d4ranc.pdf.

19 British National Archive, Nigeria, CO 323/1113/12.
20 Mary Davis, 'Labour, Race and Empire', in *The British Labour Movement and Imperialism*, eds. B. Frank, C. Horner and D. Stewart (Newcastle upon Tyne: Cambridge Scholars Publishing, 2010), 89–105.
21 British Library, India Office Archives, Use of Tear Gas in India IOR/L/PJ/12/414 [1927–1933].
22 British National Archive, Nigeria, CO 323/1113/12.
23 Ibid.
24 Ibid.
25 Ibid.
26 Martin Thomas, Violence and Colonial Order: Police, Workers and Protest in the European Colonial Empires, 1918–1940 (Cambridge: Cambridge University Press, 2012).
27 British National Archive CO 323/1341/19.
28 Ibid.
29 British Library, India Office Archives, IOR/L/PJ/12/418.
30 In 1932, Govan Brothers Ltd. set up two sugar factories in Rampur, a small Muslim state covered in farmland. Raza Sugar Co. Ltd. and the Buland Sugar Co. Ltd. leased agricultural land from Rampur, and in 1935, they partnered to form Agricultural Company, Rampur. *Govan Brothers vs Commissioner of Income-Tax*, U. P. on 10 May 1962.
31 British Library, India Office Archives IOR/L/PJ/12/418.
32 Quoted in Riaz Ahmad, 'An Aspect of the Pakistan Movement: Muslim League's Civil Disobedience Movement against the NWFP Ministry of Dr. Khan Sahib (20 February–4 June 1947)', *Pakistan Journal of History and Culture*, 28, no. 2 (2007) from NWFP Police Abstract of Intelligence Vol. 43, Peshawar, 15 April 1947, NDC Library, Islamabad.
33 British Library, India Office Archives IOR/L/PJ/12/418.
34 Ibid.
35 Ibid.
36 Ibid.
37 Ibid.
38 Bruno Latour, *Reassembling the Social: An Introduction to Actor-Network Theory* (Oxford: Oxford University Press, 2005).

Chapter 9

Controlling and Caring for Public Bodies: Civil Defence Gas Tests in World War II Britain

Alex Mankoo

> Out of an immense tear gas cloud that obliterated a large part of Union Street, Aberdeen, a tramcar suddenly materialised, the driver and passengers wearing their gas masks. Then came a telegraph messenger, his cycling speed undiminished by wearing his respirator as he threaded through the crowd.
>
> It was extremely realistic. Thousands of people, all wearing their gas masks, filled the streets from Belmont Street to Broad Street. The clouds of gas seemed to have an amazing attraction for them. Every time a bomb went off they made for the densest part of the cloud.[1]

The aforementioned might plausibly be mistaken as a script for the dystopian opening of an art-house film but is in fact a press account of a gas test held as part of a Civil Defence exercise in Aberdeen on 9 July 1941. Reading the *Aberdeen Press and Journal* description of the test from our contemporary standpoint feels unquestionably surreal, highlighting the power of context when articulating technologies – a theme that many contributors have pointed to throughout this volume. Both Anna Feigenbaum's chapter and Revill and Favera's chapter have shown how colonial contexts shaped articulations of tear gas and chemical weapons (CW) technologies. Jeanne Guillemin's analysis of how chemical warfare featured in the context of an international tribunal revealed how American interests in the chemical industry led prosecutors to obscure the role of chemicals in their articulations of harm. When reading the report of the gas test mentioned earlier, our contemporary associations of tear gas with scenes of suppression, disorder and violence conjure a context incoherent with the sense of spectacle, the mundane and the domestic in the account.

Even more surprising, then, is that this indiscriminate release of tear gas on the public was by no means a singular event. Public tear gas releases occurred in both major cities and smaller towns across the United Kingdom from February 1941. These were part of a nationwide Civil Defence effort aimed at minimizing casualties among the public in the event of enemy gas raids. Tests were typically conducted by Air Raid Precautions (ARP) wardens, and were held in public spaces such as town squares, main streets and shopping centres. While the Ministry of Home Security was responsible for Civil Defence and therefore ARP schemes, tests were not mandatory, instead being held at the approval and initiation of local authorities. Between February and June 1941 alone, at least fifty-four tear gas tests were held across Britain, with at least one in every region, including Scotland, Northern Ireland and Wales.[2] This number increased at a greater rate in the latter half of 1941, though it should be mentioned that the figure of fifty-four is derived from selected British press coverage of gas tests, such that the actual number of initial tests could have been considerably higher.

The events examined in this chapter thus serve to elucidate a period in which tear gas was articulated in a peculiar and unique fashion that stands out against the associations it has with crowd control and policing scenarios today. In these tests, tear gas featured in spaces where distinctions between domestic, military and experimental contexts were ambiguous. While this aspect is indeed similar to many contexts in which tear gas features today (e.g. scenarios where riot policing looks more like military force than domestic), what made these tests particularly unique is the way tear gas was so explicitly articulated as a technology of care. This case therefore reveals some more nuanced systems of control and experimentation associated with chemical agents, as well as the lived experiences and politics associated with the actors within these systems. ARP departments held tests with the objective of bringing 'home to the public the necessity for respirators to be carried' and creating 'in the public mind a feeling of confidence in wearing the respirator under actual gas conditions so that, in the event of the enemy using gas, people will know how to protect themselves and so avoid becoming casualties – either from gas or panic'.[3] The introduction to the 2nd edition of the Civil Defence Training Pamphlet No. 4: *Notes on Gas Tests and Exercises*, dated January 1944, read:

> Experience of Gas Tests and Exercises held in various parts of the country shows that they fulfil a highly important purpose. The public have been made more gas conscious; public confidence in respirators and in anti-gas precautions generally has been markedly increased; opportunity has been afforded to rectify any defects in respirators and respirator fitting; and Civil Defence personnel have received useful practice and instruction.[4]

This excerpt evidences the important role the realism of these tests had for the ARP authorities. This desire for realism lies noticeably in tension with the sense of excitement and attraction conveyed within the *Aberdeen Journal* account, in which people 'made for the densest part of the cloud'. In the event of an actual gas raid, this would never be the case. However, this tension between realism and excitement significantly shaped the nature of these gas tests as emergency exercises in public spaces.

Today, tear gas is commonly articulated as a technological means by which the state enforces a level of control over social movements, protests or demonstrations that seek to question the legitimacy of the state, its actions or its authority (and conversely, as a means by which such movements subvert state control). Such contexts very explicitly involve contestations over space, politics and identity but also involve attempts to order movements and populations. In his biopolitical analysis of tear gas as a weapon of governmental intervention, de Larrinaga notes, 'The deployment of teargas is also about circulation: about the ordering of movements and interactions . . . with the aim of fostering good circulation while mitigating the bad'.[5] Larrinaga refers to the conceptions of good and bad circulation set forth by Foucault,[6] in which circulation is a key instrument and target of governing processes – a sphere of operation with which populations can be secured.[7] Larrinaga suggests, 'The rationalities behind the planning of both teargas and lethal gas use is about the environment – the geography, the climate and the physical givens – and the human species and its existence in this environment'.[8] Thus, from this perspective, tear gas emerges as a means with which populations can be controlled within particular spaces, a way to create 'good' circulation (where governance is a predictable, controllable task).

Yet in the cases described in this chapter, systems of control involved a rhetoric of care that stands in marked contrast with modern narratives of (dis)order and control associated with tear gas technologies. In the gas tests, the deployment of tear gas was still about ordering and management of circulation, but what was 'good' and 'bad' was made not only explicit through notions such as proper domestic practice in wartime but also more fluid by the autonomy given to the Civil Defence services (local authorities could decide whether to host gas tests). 'Good circulation' was relatively obvious – appropriate adherence to Civil Defence protocols, for example – but the fact that local authorities remained in control of their own ARP schemes meant that gas tests varied across the geography of the nation. Local authorities were, to a degree, able to assert their own notions of control and governance.

The choice to hold tests, and the nature of the tests conducted, was therefore related to what local authorities perceived to be the best way to train and protect local populations. As such, these cases also call for an understanding of emergency service exercises as anticipatory methods of envisioning the future and ways of sorting objects and individuals, controlling behaviours, and caring for populations within particular spaces. In a study of the development of protocol in UK Fire and Rescue Service exercises,

O'Grady states, 'By their bringing future emergencies to light in the present, exercises function to assess, develop and create forms of protocol which plan out and guide . . . response to different kinds of fire emergencies, before these emergencies occur'.[9] These wartime gas tests similarly functioned as ways of bringing future gas raids into what was the present, such that civil defence practice could be assessed and refined. This simultaneously allowed the state to govern and control locales of British populations within public spaces during wartime – 'fostering good circulation' in de Larrinaga's use of Foucault's terms. In doing so, tests also operated as an informal form of large-scale human experimentation. Evaluations of these exercises were directly tied to evaluation of public behaviour, so that a successful exercise was often defined in large part by whether the public acted as anticipated and according to emergency protocols.

This chapter will be split into four parts. The next section situates these tests within the history of tear gas. Histories of chemical and biological weapons (CBW) have long noted that Britain did not use tear gas during World War II for military purposes on the European continent, nor for the purpose of crowd control (as the United States did) at home. The most oft-cited use of tear gas in Britain in this period was by police in a small number of sieges.[10] This chapter's examination of gas tests calls for a reassessment of that position, by providing a case of tear gas use that fails to fit in categories of 'military' or 'domestic' use (insofar as it is used on home soil), nor with the contemporary articulations associated with 'crowd control'. The second section will draw upon press coverage and archival documents pertaining to these tests to reveal how tear gas was framed not just as 'humane' but also as a benign technology of care. The third section explores how tests were also a form of human experimentation, in which civil defence authorities could observe and measure the social relationship local populations had with gas technologies. The final section builds upon the themes of the prior sections to demonstrate how the benign framing of tear gas, along with the experimental nature of the tests, was wrapped up in its role in a socio-technical system of state care and control. In this context, tear gas became an appropriate technology with which civil defence services could control populations within local public spaces while envisioning the sensory aspects of potential gas raids.

SITUATING GAS TESTS WITHIN THE HISTORY OF TEAR GAS

The technological trajectory of tear gas within British policy prior to World War II makes its pervasive use in gas tests all the more striking. Tear gas was first used on a widespread scale alongside choking gases (e.g. chlorine and phosgene) and blister gases (e.g. mustard gas) by major and minor powers on the battlefield and in the trenches during World War I.[11] Following this, the use of chemical weapons in warfare was condemned and prohibited through the Washington Treaty of 1922 and later the Geneva Protocol (GP) of 1925, further institutionalizing what Richard Price has called the chemical weapons taboo.[12] The latter prohibited 'the use in war of asphyxiating, poisonous or other gases, and of all analogous liquids, materials or devices' by signatory nations.[13] However, the development of this taboo also set the context for new ways of articulating tear gas, specifically ways that distinguished it from other war gases. De Larrinaga notes, 'An indication of the institutionalization of the chemical weapons taboo as well as contributing to its maintenance, both governmental programs and private industry shifted their research and development towards the production and marketing of non-lethal weapons'.[14]

It was not until this interwar period that distinctions emerged between tear gas and other chemical weapons, much of them under the rhetoric that tear gas was less lethal and more humane than other war gases. This rhetoric coincided with growing interest by both government and commercial corporations in the potential peacetime or domestic applications of tear gas. In particular, this was the case in the United States (who did not ratify the GP until 1975), where the Chemical Warfare Service (CWS) was a powerful and vocal advocate of chemical weapons, their domestic applications and associated research throughout the 1920s. Demonstrating the peacetime value of chemical weapons became increasingly necessary for the CWS as the organization suffered workforce losses and risked being disbanded after the war.[15] With tear gas as a potential lifeline, the CWS embarked on an extensive public relations (PR) campaign in support of gas as a police weapon and in 1923 was given permission to sell tear gas to the National Guard. Furthermore, the CWS provided assistance to private firms that were manufacturing, selling and marketing tear gas munitions in the United States (to police departments) – many firms that were in fact established by former members of the CWS and other gas services.[16]

Britain, however, was more reticent to adopt and develop tear gas in this period, due to what one historian characterized as a mixture of 'conservatism, political paranoia, and technical maladroitness'.[17] PR fears played a significant part, with policy makers initially wary of the possibility that the use of tear gas might be interpreted as undermining the GP. However, while Britain did not adopt tear gas for the purposes of crowd control or policing at home, it did endorse two uses: first, within the context of (secret) experimentation at Porton Down, and second, as a technology to maintain its power in the British colonies.

Within Britain, the newly founded Imperial Chemical Industries Ltd. took over a chloroacetophenone (CN) gas (a form of tear gas) manufacturing plant, set up in 1923 by another commercial company.[18] British CN output had increased twentyfold by 1927.[19] During this time, CN gas was studied extensively at the Royal Engineers Experimental Station, later renamed as the Chemical Defence Experimental Station (CDES), in Porton Down, Britain's chemical weapons defence establishment.[20] In his history of chemical weapons testing at Porton, Rob Evans notes that, between 1925 and 1936, around 4,000 human subjects were exposed to tear gases at Porton, while CN gas was also being used to test gas masks in the military and civil defence and to train the armed forces (recruits would be exposed to the gas in order to recognize it in battle).[21] With the approach and then outbreak of World War II, human experimentation at Porton intensified significantly (Porton's experimentation was by no means limited to tear gas, conducting experiments with far more toxic gases – Evans notes that tear gas tests were simply 'routine').[22]

While scientists at Porton undertook this experimentation with tear gas at home, interwar British policy makers began to consider the use of tear gas on crowds and populations in the Empire abroad. Throughout the early to mid-1930s, the Cabinet granted permission for uses of tear gas in the colonies, notably in India, Palestine and Northern Rhodesia.[23] By 1936, they authorized the colonial governments generally to purchase supplies of tear gas without prior reference to the secretary of state for 'dealing with mobs and riots in cases where it would otherwise be necessary to shoot'.[24] This decision was made on the basis that tear gas was a more humane alternative to shooting with which colonial authorities could enforce imperial control. This conception of tear gas coincided with a shift in interpretations of the GP – by the mid-1930s, many colonial policy makers no longer interpreted the use of tear gas on crowds as a contravention of the GP, a stance that both reflected the strategic interests of the British Empire and aligned with the U.S. position on the domestic use of tear gas.[25]

However, cabinet authorization did not result in the use of tear gas by colonial authorities. Anna Feigenbaum explores some of the tensions behind this in more detail with her examination of British discussions on the use of tear gas in the colonies in her chapter of this volume.[26] Colonial authorities occasionally used tear gas in the 1930s during siege situations (where individuals had sought refuge in a building, for example), but it was not until January 1939, in Burma, that they used tear gas for crowd control.[27] During World War II, use of tear gas by colonial authorities remained rare.[28]

Thus, the World War II gas tests emerged as the first instance of tear gas use on crowds within Britain, as well as the first widespread and regular usage of tear gas in any part of the British Empire. While the tests were held on the basis of strategic interests and national security – to train the public to be prepared in the event of enemy gas attack – they also represented a form of human experimentation, something that Britain was already extensively engaged in at Porton. No longer simply an object of 'humanity' with which to control colonial populations, tear gas became an object with which the local authorities 'cared' for, experimented with and trained populations on behalf of the state.

Gas tests fell under the remit of the 1939 Civil Defence Bill, which expanded upon the 1937 ARP Bill to confer additional powers upon local authorities responsible for civil defence. The Ministry of Home Security (within the Home Office) coordinated civil defence as a whole, directing central and regional services as well as local authorities. This included approving ARP schemes and coordination and supervision of the Air Raid Warden Service. In April 1939, the Ministry of Home Security appointed twelve regional commissioners for twelve regional offices, each of which coordinated the services of local authorities within their regions. While the Home Office encouraged the tests, it was not mandatory for local ARP authorities to undertake them. The decision to hold tests was instead left to the discretion of local scheme-making authorities.[29] Public gas tests did not begin until mid-February 1941 (at least, they were not publicized in the press), following a Home Office request that tests be carried out earlier in the year.[30]

The Civil Defence Training Pamphlet defined 'gas tests' as 'tear gas release schemes designed chiefly to instruct the public, give confidence in the civilian respirator and afford practice to wardens'. However, it also made note that gas tests could be combined with 'gas exercises', which involved 'tear' and 'blister' gases, or substitutes for them, and were planned mainly for the training and testing of the Civil Defence Services.[31] Both gas tests and test-exercises therefore involved members of the public who had not given consent to tests; however, combined test-exercises involved the participation of explicitly military forces as well. For example, exercises would often take the form of 'mock invasions' that involved the Home Guard and the military, as well as the police and local ARP authorities. This was the case for a large-scale exercise held in Birmingham on 9 August 1941, and an exercise in Hull on 30 August 1941 involving 'regular troops, Home Guards and the Civil Defence services'.[32] These combined test-exercises, which were not uncommon, stand out as particularly notable cases in which the distinctions between military and civilian activity blurred – the *Birmingham Post* even referred to the CAP tear gas used as a 'war gas'. The following section uses press coverage of a number of gas tests that took place in 1941 to show how tests provided a domestic backdrop in which tear gas was framed as a technology of care.

DOMESTIC CONTEXTS, BENIGN TECHNOLOGIES

> Tear gas in a mild form was released outside a cinema during a public gas test at Brighton today. Loud-speakers had given general warning during the

> previous two days that there was to be a gas alert and that the public should carry their gas-masks and wear them during the alert. Police and A.R.P. wardens stopped people approaching the tear gas area. All who had gas masks on escaped the effect but one or two of the unwary were caught . . . . There were no real casualties, but a milkman had to continue his round weeping copiously.
>
> The general test was very complete, and more of the general public wore gas masks than on any previous occasion. Bus drivers and conductors worked in their masks, as did the staffs of business houses and offices in the town.[33]

So read the *Liverpool Evening Express* account of the first British tear gas test in the United Kingdom, which took place in Brighton on 17 February 1941. Much like the *Aberdeen Journal* excerpt at the introduction of this chapter, this description situated everyday domestic activities within these tests – a milkman continuing his round, bus drivers and conductors at work, staff in the town offices going about business as usual and so on. The test featured as one of the more benign (and controllable) spectacles of everyday wartime routine, which saw 'normal' life continuously disrupted in a range of ways, such as through rationing, other air raid and civil defence precautions and actual air raids. Newspapers often covered gas tests with the language of spectacle, excitement and benevolence, and even those suffering the effects of tear gas, like the milkman, were framed as temporary, 'imagined' casualties. The *Daily Mirror*, for example, wrote of the Brighton test:

> It caused a good deal of excitement, but everywhere people co-operated with the authorities and welcomed the experiment. 'It's given me confidence in my gas-mask', [people] said afterwards. Even men and women caught without gas-masks who soon had tears streaming down their cheeks showed no resentment'. More specifically, it reported that Mrs A. Barker, of Richmond-street, a shopkeeper, said: 'People without masks came into my shop with tears running down their faces, but they all took it in good part'. Nonetheless, The Mayor of Brighton, Alderman J. Talbot told the *Daily Mirror*: 'The test was a success. But I am disappointed with the response of the rest of the town to our appeal that everyone should be carrying their masks'.[34]

Instead of describing the unpleasant effects of tear gas, the account stated that tear gassed individuals 'show no resentment' and 'take it in good part', and in contrast praised the positive, confidence building and even benevolent character of the tests. Likewise, of the effects of tear gas on a young boy during a gas test in Kingston, the *Liverpool Evening Express* remarked: 'The only "casualty" was a boy of ten who took his mask off to try the effects of tear gas. Treatment soon put him right'.[35] Notably, the word 'casualty' was placed in quotation marks. ARP wardens used tear gas to imagine casualties of possible future gas raids in the present. As such, the boy in his tear gassed state was not the actual 'casualty'. The 'casualty' referred to was the possible future state represented by the effects of tear gas on him in the present. In this scenario, tear gas was a technology of care in a potential future, such that its deleterious effects in the present were reinterpreted as benevolent. Thus, these accounts extended past the narratives of humanity and non-lethality associated with tear gas in the interwar period, to portray the effects of tear gas as benign, beneficial and protective. In some cases, its effect was even seen as humorous – of a test in the North East city of Jarrow, the *Newcastle Evening Chronicle* remarked:

> One adventurous spirit took off his mask, sniffed the air, and walked into the gas cloud. When nothing happened [to] him – he did not even shed a tear – the rest of the onlookers took off their respirators. 'It is not as strong as the curate's egg', said one of the crowd. Everybody laughed. . . . Children romped about in the gas cloud without their masks.[36]

This particular test was indeed a town spectacle, as the report claimed that 'everybody in Jarrow – 27,000 people' – carried their respirators out in the streets in preparation for the event. Gas tests may have become more routine practice for those on the home front over time, but they were nonetheless events for local communities. The domestic character of these gas tests, which paradoxically was contingent on Britain's wartime defensive strategy (preparing and protecting the population against enemy air raids), was produced by an assemblage of social values and practices, British geopolitical interests and spatial and temporal elements of both materiality and sensory experience. The legitimacy of the tests as domestic events was derived from Britain's status of being at war. For example, the domestic space of the high street, shopping centre or cinema was distinct from the military spaces of battlefields abroad but also envisioned as a potential future space of military action. The temporally transient nature of tear gas, linked with its humanity and domestic applicability, was perceived as a benign way to envision potential future use of gas and other chemical weapons by Germany (and the associated sensory experience), despite the fact that the GP prohibited the use of gas as a means of warfare. Thus, the use of tear gas for civil defence relied in part on its distinction from more lethal forms of CW, which Britain feared Germany might use, a distinction not unlike its demarcation from lethal force by colonial authorities. The bodies of those on the Home Front – many of which were women and children, framed as performing domestic roles – became human subjects through which the effect of gas raids on future bodies might be anticipated and mitigated.

GAS TESTS AS HUMAN EXPERIMENTATION

Civil Defence authorities were particularly interested in the behaviour of the British public in tests, and the effect on tear gas on them. An account of the aforementioned gas test in Kingston in the *Times* read as follows:

> Concentration of gas was laid outside a large department store . . . it was a test for the anti-gas precautions in the district affected and the ARP services in the store, while the effects of the demonstration on customers in the store itself were specially noted. . . . The public response to the appeal of the authorities for cooperation was admirable. Gas-masks were put on whenever they were required, and the injunction not to take cover or to attempt to get into shelters before fitting the masks was strictly obeyed. People carrying parcels obeyed the instruction to place them on the ground where they stood and then to put on their respirators.[37]

Thus, the tests involved forms of chemical experimentation with human subjects, but not necessarily of the type occurring at Porton Down. Rather than focusing on measuring the physiological effects of tear gas on bodies, these gas tests functioned as methods of monitoring and modifying social practices in wartime. In this case, ARP authorities observed the effects of the demonstration on customers, and monitored their response to injunctions. As such, tests also functioned as a measure of control and order. Authorities measured the extent to which the public obeyed official injunctions such as placing parcels on the ground, and not taking cover, before adorning respirators. Sometimes measurements were quantified, and if not they were certainly monitored through observation. For example, following a test in Birmingham on 13 August 1941, Birmingham ARP Committee Chairman Norman Tiptaft noted, 'Now as to the necessity for this exercise. A census taken on the morning of August 13 showed . . . only one person in thirty four was carrying a gas mask. The next morning (August 14) one person in seven – just about five times as many – was carrying a gas mask'.[38]

In this respect, these tests functioned as experiments on human subjects' relationships with gas technologies (and air raid technologies), such as tear gas, warning gas rattles and gas masks. Authorities wanted to measure wartime behaviour, responses to gas, and the way in which both wardens and the general public handled gas tests, while also monitoring gas masks and the effects of tear gas. Gabriel Moshenska has previously highlighted how gas vans and school gas chambers used tear gas to educate the public in the value and efficacy of their gas masks.[39] (The gas vans and chambers Moshenska discusses were another part of anti-gas training under the remit of civil defence personnel. They operated as chambers within which civil defence personnel and members of the public could test their respirators in an atmosphere of irritant gas, often CAP tear gas. This chapter, however, maintains its focus on the more indiscriminate tear gas tests that were held in public spaces.) Summarizing the outcomes of the Kingston test, the *Times* noted:

> A satisfactory feature of the tests was their effect on the local inhabitants, who took care to have their respirators ready for use. In one way the tests proved inconclusive, for a strong wind dissipated the gas as soon as it was released; but valuable information was obtained about the behaviour of the public and the reactions of the local ARP workers. The objective of the first exercise was to find out how quickly gas could be detected by patrolling wardens and how wardens in the neighbouring post areas would act.[40]

The claim that 'valuable information was obtained' about the behaviour of human actors despite the fact that 'in one way the tests proved inconclusive' (due to the wind's effect on the gas) again illustrates how the relationship between human actors and gas technologies became an experimental focus. However, contrary to the upbeat tone adopted in many newspaper reports, some members of the British public were understandably disgruntled about the use of tear gas in tests, for a range of reasons. Ada Croxton of Stourbridge complained that tests were unnecessary in a letter to the *Evening Despatch*: 'There are already enough poisons in the air without endangering precious eyesight by releasing tear-gas in the streets. I hate gas-masks, and the type of people who always carry them. I don't think there is the slightest need for us to fear a gas attack, therefore I will not carry a gas-mask'.[41] Another person raised concerns for vulnerable people subjected to gas, specifically infants and individuals with breathing conditions, and noted the difficulties the tests posed for mothers. They also pointed out the dangers blinded pedestrians faced from vehicles.[42] Clearly, some individuals felt unhappy with tear gas being used against them without their consent. Yet mention of consent was notably absent in planning gas tests.

Schmidt has shown how pressures of secrecy, national security and the emergency state of being 'at war' justified the suspension of ethical principles for those involved with programmes of human experimentation at Porton Down.[43] While gas tests were not formal experiments, they were, like experimentation at Porton, a result of national security pressures and the state of wartime emergency. They too involved extensive use of chemical agents on individuals who had given no consent, and they too incorporated kinds of observation and measurement. Unlike what went on at Porton, however, the primary aim of the tests was not to gain knowledge of tear gas chemicals. Rather, tests were to observe and control social responses to, and understandings of, the technology. Brian Balmer has also elsewhere contributed to literature on more informal and impromptu programmes of human experimentation. Most pertinent is his work on the case of the *Carella Trawler*, a fishing vessel that strayed into the danger zone after a British biological test bomb detonation off the Isle of Lewis in 1952. The authorities holding the trial chose not to notify the crew and instead tailed and monitored the trailer for a month, listening for a distress call while the navy developed contingency plans in case the crew became seriously ill.[44] Like the gas tests, this was both a national security operation (discretion about the biological weapons (BW) programme was paramount) and a form of human experimentation (monitoring the effects of the test on the crew).

However, unlike the experimentation at Porton, accidents like the Carella case, and contemporary riot control scenarios, where target populations might be divided into categories such as 'rioters', 'bystanders' and 'subjects', these tests were instead intentionally indiscriminate. The intended targets of the gas were any, and all, individuals passing through the spaces within which the gas permeated, such that the precise individuals affected would be unknowable in advance (and perhaps even after). For civil defence authorities, targeting the population in this way is part of what lent realism to the exercises. In this respect, acquiring consent from those affected by the tests was an impossible task. Newspapers did, however, often feature official warnings in advance of tests to inform readers of their occurrence, in which local authorities chose whether to disclose details such as time and place. Some warnings specified the exact times and places of tests, whereas others intentionally left this vague to encourage public vigilance, simply noting that a test would happen in the local vicinity at an undisclosed time.

Balmer has also examined the rationales behind large area germ warfare tests in the UK in the 1960s and 1970s, in which populations were sprayed with bacterial suspensions of *E. coli* and *B. globigii* (two spores thought to be harmless) in trials designed to assess the biological weapons threat to national population.[45] The reasoning behind World War II tear gas tests was similar to what Balmer identifies as the principle behind these large-scale tests: using population bodies (the populations being sprayed) to protect the national body. The use of chemical and biological agents was, from the perspective of government, military, and civil defence, for the population's own good. Yet, in large area germ tests, populations were caught up in the experiment but not part of it in any obvious manner, such that they constituted 'implicated actors', approximating 'an artifact needing to be excluded from consideration'.[46] Balmer argues that this discourse of exclusion allowed for arguments about consent to be readily countered as the population were not part of the trial and consent was therefore unnecessary. In contrast, as already mentioned, the gas tests in World War II benefited from as many people being involved within the spaces of the test as possible, such that the opposite effect occurred. Because tests were designed to reach everybody within the space, acquiring consent was neither required nor feasible. Traditional experimental notions of 'subject' and 'experimenter' did not apply because these were relatively flexible categories within the space of the test.

Consent may have been absent from ARP discussions, but some authorities raised concerns regarding ARP liability in the event of a member of the public being injured in a test.[47] These parties requested that the government provide them with indemnity against such action, which was given by the Treasury in September 1941 for (1) cases where damages were awarded by the Court against an authority, or (2) where a local authority agreed to pay a member of the public without the case coming before the court (provided the sum was reasonable and approved by the regional officer).[48] This admission of indemnity indicates that authorities were aware of the potential risks of the tests but perceived these risks to be acceptable. Moreover, given the public nature of the tests, the question of consent was seemingly irrelevant – compensation was to be granted to injured individuals, whether or not they endorsed the tests. Again, this points to the function of the tests as ways of envisioning future emergencies. The risks of injuries to the public during tests were acceptable in light of the potential risks to the public from enemy gas raids in the future. Notions of care and security were predicated on anticipated temporal states, such that they legitimated the gas tests in the present. This aspect of the tests will be examined in the final section.

FUTURE EMERGENCIES, PRESENT EXPERIENCES

As previously mentioned, a key function of the gas tests was to provide training to the Civil Defence services. The Civil Defence Training Pamphlet (for ARP wardens and authorities) noted:

> Whilst it is necessary, in view of the possibility of gas attacks, that all scheme-making authorities should push forward with their plans for holding gas tests and exercises, it is equally necessary that:-
>
> a) All gas tests and exercises should be carefully planned with due attention to realism, and conducted on sound, practical lines.
> b) The lessons which the exercises are intended to teach should be clearly defined and properly learnt both by the personnel taking part, by those standing by and by the general public.
> c) Personnel taking part, particularly at tests and exercises including the public, should have sufficient anti-gas knowledge to enable them to carry out their duties at the exercises without making mistakes which will be obvious to the public.
> d) Prior to the incident all personnel should be carrying on normally at their usual place of duty. There is no realism about a test or exercise when squads

> of men in full protective clothing are kept ready waiting for the gas warning to sound.[49]

The pamphlet made clear that authorities saw the realism of the tests as imperative. This realism enabled authorities to simulate the experience of possible future emergencies, allowing them to refine protocols and train the public and defence workers in civil defence standards. The realism was directly associated with the use of tear gas, which became a way of experiencing the physiological, visual and emotive character of potential gas raids. In the case of the Fire and Rescue Service, O'Grady notes 'The capability of exercises to offer realistic impression of future emergencies is intrinsically tied to the question of what material props are found in the exercise'.[50] In ARP gas tests, gas masks, tear gas technologies, protective clothing and gas rattles (to signify the beginning of gas raids) were examples of such material props. These facets of the tests also speak to a notion that they were a form of simulation, generating 'knowledge of gaps, misconnections and unfulfilled needs', in the words of Lakoff.[51] This is exhibited in point (b) in the Civil Defence Pamphlet: 'The lessons by which the exercises are intended to teach should be clearly defined and properly learnt both by the personnel taking part'. The tests therefore made 'infrastructural vulnerabilities visible' and entailed a method 'for designating priorities and allocating resources in a preparedness system'.[52] Making such vulnerabilities visible was thus a way of creating 'good circulation' in wartime within local public spaces. Balmer, Spelling and McLeish expound a similar conception of simulation regarding the use of tear gas in their contribution to this volume.

Similarly, the request (c) that personnel carry out exercises without making mistakes noticeable to the public highlights how successful performance of the tests hinged upon a public experience that felt 'real'. The sight of personnel in protective clothing prior to the test, for example, would be a signifier of a test's superficiality and was to be avoided. Tests were a means of building public confidence not just in equipment but the entire socio-technical system – ARP wardens, technologies, emergency protocols – of civil gas defence. Like Fire and Rescue Service exercises, the performance of these tests relied on 'invoking different emotional states in participants through the presence of material props which entangle with the discursive and audio-visual invocation of the future emergency'.[53] Furthermore, the realism of the performance was associated with the accuracy of the gas tests as experiments. The pamphlet point (d) made clear that a counterfeit performance would not provide civil defence services with reliable experience of how the British public would respond in the event of an actual gas raid. The 'results' of the test as experiment were therefore also contingent on its realism. The bodies of the British public themselves became material props through which the effect of gas raids on future bodies could be anticipated and mitigated.

Yet, as shown in an earlier section of this chapter, the press coverage often paradoxically portrayed gas tests as exciting, even humorous and fun events. A real gas raid would not be any of these things. This reading was enabled through the sensory effects of tear gas. British policy makers had already interpreted tear gas as 'humane' when authorizing it in the colonies, on the grounds that its effects were temporary and provided an alternative to shooting. In the tests, the temporary effects of tear gas served to at once provide a sensory experience of gas in an emergency that was 'real' enough to simulate enemy use of war gas, while remaining short-lived enough that they only did so within the temporal limits of the exercises. In doing this, tear gas was constructed not just as humane but also as a technology of care within an assemblage of values, technologies, social actors and practices that in turn provided legitimacy to the gas tests as programmes of state control and care. For instance, exercises included gas cleansing centres where individuals affected by gas could be decontaminated during tests (and issued clothing to go home in). Local authorities assigned decontamination squads to the task of monitoring and decontaminating both people and spaces following a test (this included, e.g., decontamination of food in stalls).[54] Through these assemblages, ARP authorities were able to control and secure locales of heterogeneous British publics across the nation on behalf of the state. Like the large-area germ tests examined by Balmer, tests on local population bodies became a way of protecting the national body.

This further speaks to de Larrinaga's notion of tear gas as an apparatus of security that fosters good circulation and mitigates the bad. Extrapolating this to the case of gas tests, this good circulation consisted of, most notably, the carrying and wearing of gas masks. In World War I, tear gas was often employed on battlefields as a way of forcing the enemy to wear gas masks, lessening their efficiency.[55] In gas tests, tear gas had a similar effect but did so in the context of a programme of state protection, forcing civilian populations to practice wearing gas masks in preparation for enemy attack. Gas tests also allowed for the identification and marginalization of bad circulation. Following the first test in Brighton, the *Portsmouth Evening News* remarked, 'To go without the gas mask is to play into the enemy's hands by helping the aggressors and handicapping the defence'.[56] Similarly, the *Daily Herald* featured two images from a gas test in Southend. The first image showed a woman and child wearing gas masks with the heading 'They Were Ready' and caption 'This mother and child were prepared . . . they suffered no effects, but look at the picture of the girl below'.[57] Below this was a picture of a woman suffering the effects of gas, with the heading 'But She Wasn't' and the caption 'Caught without her gas-mask, she was soon "crying" as the tear-gas took effect. So wherever you go, carry your gas mask'. Local civil defence authorities issued gas masks, and everyone was expected to have one. Gas tests were thus a way of testing this, and identifying individuals who required, or had forgotten, their masks.

This chapter offers a case in which tear gas was used as a means of controlling populations, within tests and exercises designed as a means for envisioning and enacting imagined future states of emergency. It has shown how the temporary effects of tear gas allowed for a reading of it as not only a humane technology but also a technology of care and protection. Furthermore, it also suggests that gas tests represented an informal programme of human experimentation, in which authorities observed and measured human subjects' relationships with gas technologies as opposed to only the physiological effects of tear gas on those bodies.

Tests blurred the lines between the civilian, military and experimental. While they were usually represented as domestic events, they could involve military actors, and were necessitated by Britain's being at war. To describe the tests as a form of 'crowd control' seems deficient; however, they do call for a reassessment of the position that Britain did not use tear gas on crowds at home until well after World War II. Gas tests were a form of 'crowd control' insofar that they were attempts to control, order and sort populations, through training them in particular civil defence practices.

The tests are perhaps not comparable to the police use of tear gas and riot control agents (RCAs) today, but they do highlight a historical case in which the boundaries between military and domestic action were traversed. In recent years, scholars have been increasingly concerned over the use of tear gas and other non-lethal weapons (NLWs) in paramilitary operations, noting the potential this has for the erosion of the international norm against chemical and biological weapons.[58] At such a time, it is all the more important that these forgotten past cases of military-domestic gas use come to light, particularly for the implications that they might have going forward for understanding the historical trajectory of tear gas technology.

NOTES

1 'Aberdeen under Gas and H.-E. "Bombs"', *Aberdeen Press and Journal*, 9 July 1941, 3.

2 In 1939, the Ministry of Home Security divided mainland Britain into twelve Civil Defence regions (the headquarters of which were Newcastle, Leeds, Nottingham, Cambridge, London, Reading, Bristol, Cardiff, Birmingham, Manchester, Edinburgh and Tunbridge Wells). Civil Defence in Northern Ireland was directed from the central office in Belfast.

3 The National Archives [TNA], HO 186/481, 'Public Tear Gas Exercises, Liability of Local Authorities'; TNA, HO 186/481, Reference 722, 298/29, From Brown, 22 May 1941.

4 TNA, HO 186/1575, Civil Defence Training Pamphlet No. 4, 2nd ed., *Notes on Gas Tests and Exercises*, Issued by the Ministry of Home Security (London: HMSO, January 1944).

5 Miguel de Larrinaga, '(Non)-Lethality and War: Tear Gas as a Weapon of Governmental Intervention', *Critical Studies on Terrorism*, 9, no. 3 (1 September 2016), 529.

6 Michel Foucault, *Security, Territory, Population* (Basingstoke: Palgrave, 2007), 18.

7 Claudia Aradau and Tobias Blanke, 'Governing Circulation: A Critique of the Biopolitics of Security', in *Security and Global Governmentality: Globalization, Governance and the State*, ed. Miguel de Larrinaga and Marc G. Doucet (London: Routledge, 2010).

8 Larrinaga, '(Non)-Lethality and War', 529.

9 Nathaniel O'Grady, 'Protocol and the Post-Human Performativity of Security Techniques', *Cultural Geographies*, 23, no. 3 (1 July 2016), 495.

10 Mike Waldren, *Tear Gas and Empire* (Chatteris: Police Firearms Officers Association, 2013).

11 SIPRI, *The Problem of Chemical and Biological Warfare Volume I: The Rise of CB Weapons* (New York: Humanities Press, 1971).

12 Richard MacKay Price, *The Chemical Weapons Taboo* (London: Cornell University Press, 1997).

13 United Nations, Protocol for the Prohibition of the Use in War of Asphyxiating, Poisonous or Other Gases, and of Bacteriological Methods of Warfare, 17 June 1925, available at http://www.refworld.org/docid/4a54bc07d.html.

14 Larrinaga, '(Non)-Lethality and War', 530.

15 Thomas Faith, '"As Is Proper in Republican Form of Government": Selling Chemical Warfare to Americans in the 1920s', *Federal History*, no. 2 (2010), 28–41.

16 Daniel P. Jones, 'From Military to Civilian Technology: The Introduction of Tear Gas for Civil Riot Control', *Technology and Culture*, 19, no. 2 (1 April 1978), 151–168.

17 Simeon Shoul, 'British Tear Gas Doctrine between the World Wars', *War in History*, 15, no. 2 (1 April 2008), 168–190.

18 See Robert Evans, *Gassed: Behind the Scenes at Porton Down* (London: House of Stratus, 2000), 437. Imperial Chemical Industries Ltd. was founded in 1926 and later became the largest chemical manufacturer in Britain.

19 Evans, *Gassed*, 264.

20 From 1916 to 1929, this was named the Royal Engineers Experimental Station; from 1929 to 1930, the title changed to the Chemical Warfare Experimental Station; then from 1930 to 1948 to the Chemical Defence Experimental Station; from 1948 to 1970 to the Chemical Defence Experimental Establishment; from 1970 to 1991 to the Chemical Defence Establishment; and from 1991 to 1995 to the Chemical and Biological Defence Establishment (CBDE) before being merged into the Defence Evaluation and Research Agency in 1995. Today, the CBDE falls under the Ministry of Defence, within the Defence Science and Technology Laboratory (DSTL).

21 Evans, *Gassed*, 264.

22 Ulf Schmidt, *Secret Science: A Century of Poison Warfare and Human Experiments* (Oxford: Oxford University Press, 2015); Evans, *Gassed*.

23 See Waldren, 'Tear Gas and Empire'; Edward M. Spiers, *Chemical Warfare* (Chatteris: University of Illinois Press, 1986); Alex Mankoo, 'Teargas – We Haven't Got the Foggiest: Deconstructing the Ambiguities of Creeping Legitimisation' (MSci diss., University College London, 2011); Anna Feigenbaum, *Tear Gas: From the Battlefields of WWI to the Streets of Today* (London; Brooklyn, NY: Verso Books, 2017).

24 TNA, CAB 24/257/46, 'Cabinet. Use of Tear Gas in the Colonial Empire', December 1935.

25 Feigenbaum, *Tear Gas*; Mankoo, 'Teargas'; Waldren, 'Tear Gas and Empire'.

26 See also her recent book: Feigenbaum, *Tear Gas*.

27 Shoul, 'British Tear Gas Doctrine between the World Wars', 189; Britain did, however, aid Allied governments in dealing with crowd control – Ulf Schmidt points out that Porton supplied South Africa with bombs filled with tear gas that the South African government used against opposition groups. See Schmidt, *Secret Science*, 60.

28 Waldren describes an incident in 1942 where Bombay police used tear gas to break up a mob (Waldren, 'Tear Gas and Empire', 17).

29 TNA, HO 186/481, 'Public Tear Gas Exercises. Liability of Local Authorities'.

30 This request is mentioned in the *Times*. See 'Home News. Tear Gas among Shoppers', *The Times*, 1 April 1941, 2. This request may have come following Churchill's national broadcast on 9 February 1941, in which he declared, 'We must all be prepared to meet gas attacks, parachute attacks and glider attacks, with constancy, forethought and practiced skill'. See 'Give Us the Tools, and We'll Finish the Job', Speech Broadcast by Prime Minister Winston Churchill, 9 February 1941, available at http://www.ibiblio.org/pha/timeline/410209awp.html.

31 TNA, Civil Defence Training Pamphlet No. 4.

32 'Gas Exercise in Birmingham. Citizens' Behaviour under Attack. Large-Scale Defence Scheme', *The Birmingham Post*, 11 August 1941. Tear gas was only used in one area of Birmingham in this exercise; 'Home Guards Took the Honours: Mock Invasion Test of Hull Defences', *Hull Daily Mail*, 2 September 1941.

33 'Town Stages Tear Gas Surprise Test', *Liverpool Evening Express*, 17 February 1941.

34 'Trust Mask after Tear Gas "Raid"', *Daily Mirror*, 18 February 1941.

35 'Tear Gas Test in Streets', *Liverpool Evening Express*, 31 March 1941, 1.

36 'Gas Exercise but Everybody Laughed', *Newcastle Evening Chronicle*, 19 April 1941, 5.

37 'Home News. Tear Gas among Shoppers', *The Times*, 1 April 1941, 2.

38 'Tear Gas. More Experiments in Birmingham. A.R.P. Chief's Advice', *The Birmingham Mail*, 20 August 1941, 4.

39 Gabriel Moshenska, 'Government Gas Vans and School Gas Chambers: Preparedness and Paranoia in Britain, 1936–1941', *Medicine, Conflict and Survival*, 26, no. 3 (1 July 2010), 223–234.

40 'Home News. Tear Gas among Shoppers', *The Times*, 1 April 1941, 2.

41 'Public Opinion. Tear Gas', *Evening Despatch*, 14 August 1941, 3.
42 'Gas Test without Warning', *The Rochdale Observer*, 12 April 1941, 7.
43 Schmidt, *Secret Science*, Chapter 3.
44 Brian Balmer, 'How Does an Accident Become an Experiment? Secret Science and the Exposure of the Public to Biological Warfare Agents', *Science as Culture*, 13, no. 2 (June 1, 2004), 197–228; Brian Balmer, *Secrecy and Science* (Burlington, VT: Ashgate, 2012), Chapter 3.
45 Brian Balmer, 'Using the Population Body to Protect the National Body: Germ Warfare Tests in the UK after WWII', in *Useful Bodies: Humans in the Service of Medical Science in the Twentieth* Century, eds. J. Goodman, A. McElligott and L. Marks (John Hopkins Press: Baltimore, 2003), 27–52.
46 *Ibid.* Balmer uses the term 'implicated actors' in reference to M. J. Casper and A. E. Clarke, 'Making the Pap Smear the "Right Tool" for the Job', *Social Studies of Science*, 28 (1998), 255–290.
47 TNA, HO 186/481, From J. R. Alderson to L. N. Helsby, 13 September 1941.
48 TNA, HO 186/481, From L. N. Helsby, Treasury Chambers, 22 September 1941.
49 TNA, Civil Defence Training Pamphlet No. 4.
50 O'Grady, 'Protocol and the Post-Human Performativity of Security Techniques', 500.
51 Andrew Lakoff, 'Preparing for the Next Emergency', *Public Culture*, 19 (1 April 2007), 247–271.
52 Ibid., 266.
53 O'Grady, 'Protocol and the Post-Human Performativity of Security Techniques', 496.
54 TNA, Civil Defence Training Pamphlet No. 4.
55 Jones, 'From Military to Civilian Technology', 152.
56 'Respirator Test', *Portsmouth Evening News*, 18 February 1941.
57 'They Were Ready', *Daily Herald*, 31 March 1941, 3.
58 Kai Ilchmann and James Revill, 'Chemical and Biological Weapons in the "New Wars"', *Science and Engineering Ethics*, 20, no. 3 (September 2014), 753–767; Julian Perry Robinson, 'Difficulties Facing the Chemical Weapons Convention', *International Affairs*, 84, no. 2 (2008), 223–239.

Chapter 10

'Chemical Bodies' and the Future of Control

Alex Mankoo and Brian Rappert

The chapters in this volume have explored many ways in which chemical agents both reconfigure and are reconfigured by bodies, and how these relations constitute forms of ordering, control and mess. 'Chemical bodies', the title of this volume, reflects this co-constitutive relationship.[1] What a chemical agent 'is' in a given context emerges with how it is situated in relation to individuals, environments, institutions and expertise; thus, it is produced in tandem with particular normative values and ideals that shape the role it then performs in that context. Thus, attempts to order and control chemicals are at once attempts to order and control bodies; knowledge about chemicals is contingent upon knowledge about bodies, and vice versa. The contributions here have enquired into a range of such configurations, where chemical agents are situated within heterogeneous assemblages of objects, individuals and groups. In doing so, they have brought into focus a variety of 'bodies' – bodies of individuals, bodies of collectives (the body of a crowd, population bodies, bodies of 'in-groups' and 'out-groups'), bodies of knowledge and institutional bodies.

Thinking with this notion of 'chemical bodies' therefore calls for attending to forms of power and attempts at control produced by specific configurations. This attention consequently reveals tensions and similarities between configurations – for example, how using medical bodies of knowledge to construct a chemical as a drug, as the Himsworth Committee did with 2-chlorobenzalmalononitrile (CS) gas, lies in tension with how CS might be used in practice by police, or with particular effects it might have on individuals within the context of a crowd. In this process, sometimes certain bodies are foregrounded, while others are ignored. These movements are matters of techno-politics through and through. In the International Military Tribunal for the Far East (IMTFE), for instance, broader geopolitical interests entangled with institutional pressures from the Chemical Warfare Service (CWS) and Supreme Commander for the Allied Powers (SCAP) to lead the International Prosecution Section (IPS) to leave the charges of Japanese use of chemical weapons (CW) in the tribunal unspecified and vague. This in turn both concealed the technological capacity of chemical agents to harm as well as the violence and injuries the agents specifically inflicted upon many Chinese victims.

At other times, however, certain instances of chemical and biological weapons (CBW) use take the limelight of attention above other forms of killing and injuring. This was certainly the case during the professional hearing against Wouter Basson, when prosecutors focused on allegations surrounding the use of tear gas over the development of an anti-fertility vaccine because this was seen to be legally expedient. Similarly, the use of tear gas in British civil defence tests during World War II was a notable point of focus for civil defence officials, the public and the media. However, its use was situated within contexts of civil defence practices rather than police or military application, and in this context, it was deemed a means of *protecting* the bodies of the public rather than attacking them. Furthermore, mention of the tests is distinctly absent among contemporary historical CBW literature, itself highlighting the temporal aspect of shifts in visibility.

These chemical body configurations therefore involve unique enactments – they are not simply sources of meaning but also of practice. To use the words of Annemarie Mol in her book *The Body Multiple*, 'ontologies are brought into being, sustained, or allowed to wither away in common, day-to-day sociomaterial practices', in which 'each event . . . turns some "body" (some disease [some chemical agent], some patient [some subject]) into a lived reality'.[2] Chemical bodies are, too, multiple – they too are brought into being.

These ways in which chemical bodies become constituted and re-constituted also return us to the overarching themes of distinction, articulation and experimentation outlined in the introductory chapter. If distinctions cordon off parts of the world from each other, articulations entail deciding which distinctions should come to matter, and enacting them accordingly. Experimentation, then, is an exercise in managing tensions: it involves stepping away from taken-for-granted distinctions to ask how things might 'actually' exist while at the same time adhering to certain distinctions (e.g. who or what the research subject is) as a means with which to test that idea. It involves an acknowledgement of the limits of what is known and unknown but in doing so entails a commitment to particular ideas regarding what it means to know.

The following sections revisit and tie together the volume's thematic strands of distinctions, articulations and experimentation while highlighting crossover between chapters, before this chapter closes with a discussion of the implications of our chemical bodies approach more generally.

REVISITING: DISTINCTIONS

Distinctions imply differences. Therefore, making distinctions functions as an act of categorizing one thing as different from another in some particular way. There are often a multitude of features by which to distinguish things (one might distinguish two objects according to their size, or their colour, their purpose, their location, etc.). In this respect, deciding what counts as a pertinent distinction is itself an act of ordering, framing particular debates and political issues as more or less significant. For instance, one of the more obvious ways of demarcating chemical agents is to distinguish them from conventional weaponry. This might speak more to those outside of the professional world of arms control, and to the desire to locate some singular basis (in this case, killing) by which to establish what counts as a moral transgression. Justin McBrayer, a philosopher at Fort Lewis College, Colorado, questioned this distinction in a recent commentary in the *Denver Post*:

> You might think that what's special about the use of poison is not how it kills people but that it has the potential to kill so many. On this way of looking at it, it would make sense to draw a line between the use of conventional and chemical weapons, just as we do between conventional and nuclear weapons. In other words, the distinction between conventional and chemical is really just a proxy for the morally relevant distinction between killing a few and killing the many. But there isn't a shred of evidence for this position. Current conventional weapons kill many, many times more people than your basic chemical weapon. This is true of the conflict in Syria. Sadly, the recent chemical attack claimed the lives of 80 innocent people. But this pales in comparison to the average of 257 people who die each day in Syria by conventional means. Our foreign policy should be both consistent and just. Insisting on a difference between conventional murder and chemical murder is neither.[3]

Referring to the April 2017 sarin gas attacks in northern Syria, McBrayer's commentary is another iteration of a debate that has been alive for over a century – why, and how, chemical weaponry should be demarcated from 'conventional weapons'. Early twentieth-century militarists and CW advocates such as J. F. C. Fuller and J. B. S. Haldane appealed to the distinction between killing the many and killing the few, albeit for very different reasons to the *Denver Post* commentary.[4] Those proponents used this argument to claim that chemical warfare was more humane than conventional means and as such should not be prohibited in armed conflict. In stark contrast, McBrayer's commentary is questioning the distinction between the chemical and conventional weaponry as a means of asking why foreign policy makers pay special attention to chemical acts of violence over and above conventional acts of violence.

Building upon this, rather than simply asking why, in the case of Syria, chemical killing is differentiated from conventional killing, we can go further to ask: why is that particular distinction being made at this particular time, in this particular space, with these people involved? As we have seen throughout this volume, harmful and fatal uses of chemical agents do not always result in immediate punitive action, condemnation or evaluation of policy. The deaths of thirty-seven prisoners in Egypt, or fourteen migrants in Spain, through the use of tear gas (as raised in Michael Crowley's chapter),[5] for example, resulted in debates regarding appropriate use of force (instead of comparisons between conventional and chemical weaponry). Both tragedies present symptoms of the othering highlighted in Revill and Favera's chapter – prisoners and migrants were perceived as 'outgroups' to the extent that chemical violence upon them was considered warranted by those in power. Furthermore, many of the contributions in this volume demonstrate how geopolitical contexts transform bodies and the meaning behind the use of chemicals upon them. Dismissing the distinction between conventional and chemical killing on the moral basis of instead focusing on the extent of killing thus does not account for the distinctions drawn around what counts as 'killing', who can be 'killed' and how 'killing' can be known. Nor does it recognize the contextual elements (actors, institutions, affect, materiality, etc.) that might be associated with chemical rather than conventional force. In this respect, distinctions also operate as a means of marking which bodies matter, and which do not – whose bodies are human and whose bodies are in a real sense 'less than human'.

Jeanne Guillemin's chapter on the Japanese chemical warfare programme in the mid-twentieth century identified how the effects of chemicals on victim's bodies were obscured and erased during the 1946–1948 IMTFE. The distinctions between what and who did and did not matter in relation to the tribunal were negotiated in large part according to the U.S. military's interest in preserving the CWS. Moreover, Guillemin pointed out how distinctions in the IMTFE briefs' drafts shifted over time – for example, chemical warfare charges in relation to the Geneva Protocol (GP) disappeared between the 13 May 1946 and 24 May 1946 drafts. The 24 May draft made a distinction between the use of vesicant gases and tear, sneezing and vomiting gases, categorizing gas use as for emergency purposes rather than chemical warfare, and emphasized the distinction between atomic warfare and chemical warfare. These distinctions downplayed Japan's use of chemical weapons, simultaneously rendering the bodies of victims less visible. Omitting the Chinese charges of CW use against Japan, the IPS effectively ordered potential harm to international (military) geopolitical interests above the harms to the bodies of the victims. This was executed in part through a shifting of the boundary of what counted as 'chemical warfare', displaying a moment of 'co-production' (where what counted as CW was co-produced with notions of the institutional mechanisms that 'ought' to govern its use).[6]

In terms of how contemporary distinctions are being played out, the 2007 issue of the UK Ministry of Defence (MoD) army field manual on Counter Insurgency Operations (Strategic and Operational Guidelines) identified 'calmative agents and irritants' as well as 'incapacitants' as non-lethal weapons (NLWs) that might 'represent additional capabilities for use in military commanders' graduated use of force to deter, defend or attack an opponent'.[7] It went on to say, 'Research and development of a wide range of techniques in the use of non-lethal weapons (NLW) indicates that this will become a major factor in future counter insurgency campaigns'.[8] Some sections of the manual point more explicitly to notions of care and protection being mobilized as justifications for NLW: 'Public concern for losses among the belligerents and civilian population and the associated requirement for the minimal use of force have increased interest in the potential for NLW, especially in peacekeeping and counter insurgency operations'.[9]

This argument for NLW on the basis that they constitute 'minimal use of force' is notably similar to the arguments colonial policy makers made for the use of tear gas as a means of control (as explored in Anna Feigenbaum's chapter in this volume). Much like the rationale upon which the British Cabinet authorized the use of tear gas in the colonies in 1936, the 2007 MoD field manual states, 'What is new and has enhanced the importance of NLW is the prevailing security climate in which the use of force, especially for UN operations, has become almost commonplace. Until now troops on peacekeeping operations have been authorized to use lethal weapons only in self-defence. If forced to open fire, they shoot to kill'.[10] Once again, here the value of NLW comes from comparisons to the use of lethal force in contemporary peacekeeping scenarios.

In contrast, in this volume, Michael Crowley eschewed the contested framing term 'non-lethal weapon' and instead explores and compares how two distinct categories of 'weapons employing toxic chemicals' – riot control agents (RCAs) and incapacitating chemical agents (ICAs) – have been developed, promoted and used to varying degrees in law enforcement. Crowley examined the ambiguities that exist regarding what these chemicals 'are', and therefore how they are controlled and governed, within the text of the Chemical Weapons Convention (CWC). He showed how this was tied to where states have (or have not) drawn distinctions between the acceptable and prohibited purposes for such chemicals in law enforcement, arguing that 'law enforcement' is often a taken-for-granted category in the CWC. As a result, RCAs have been widely misused for human rights violations by law enforcement officials, and when used in excessive amounts have proved injurious or fatal to those targeted, particularly in confined spaces.

In our analysis, we might think here of how what counts as a RCA comes hand in hand with what counts as law enforcement. Crowley notes how the legitimacy of law enforcement is related to interpretations of the 'types and quantities' of chemicals being used. He points out how, in practice, RCAs have been used in the name of law enforcement for potentially legitimate applications as broad as those 'beyond control or dispersal of rioting crowds, including for the incapacitation of violent individuals, personal defence, and in counter-terrorist operations'. With this in mind, rather than making the category of law enforcement, with its associated 'types and quantities' of permitted toxic chemicals, the means of determining what chemical use is or is not acceptable, what would come from incorporating a range of conceptions regarding what should constitute their legitimacy? For example, how might incorporating ideas of human rights law alongside, or within, the law enforcement category change the way chemical control manifests in the CWC?

In contrast to this general acceptance by states of the appropriate use of RCAs in law enforcement, Crowley noted that no such consensus exists with regard to the employment of ICA weapons. Instead, scientific and medical associations, human rights organizations and a growing number of CWC states parties are making a concerted effort to clarify that such ICA weapons are prohibited under the CWC.[11] Crowley's chapter, which called for a more contextualized, nuanced and holistic conceptualization of chemical control, thus suggested that there is not only an opportunity but also an urgent need, to apply such a conceptualization to this emerging category of ICA weapons.

As the boundaries between police and military bodies blur – which they do in many of Crowley's examples – so do the boundaries between the technologies associated with those contexts, once again reflecting the utility of our notion of 'chemical bodies'. The blurring of these distinctions was also at the crux of Mankoo's study of British tear gas tests during World War II. By placing tear gas technology within a context of civil defence practices, it was at once demarcated as a form of protection different from other war gases, but also considered alongside them as a way of evoking realism in civil defence practices.

Relatedly, Feigenbaum's chapter also drew attention to the various ways a distinction was drawn between tear gas use for the colonial and home markets in the interwar and post–World War II British Empire. Her work highlights the boundaries that existed regarding who should be gassed, but also points to how policy makers decided whether they understood tear gas to be under the remit of the GP (and how this shifted over time). This was in part linked to the distinction between lethal forms of force and non-lethal forms of force being increasingly made by the United States and Britain's colonial police forces. Feigenbaum points out how colonial authorities saw tear gas as 'a third way out' that allowed governments to save face in the international arena while still quelling resistances. Furthermore, her chapter draws out the range of epistemic frames with which the efficacy of tear gas was evaluated – for example, by its financial value, its technical efficiency (e.g. shelf life), public reaction to its use and its relevance for tasks of colonial policing – and shows how these different epistemic approaches were foregrounded or backgrounded at different times.

Balmer, Spelling and McLeish spoke even more so to how epistemic approaches can lie in tension and in harmony with one another, examining what led to the categorization of CS tear gas as a drug rather than a weapon in the Himsworth Committee report. Feigenbaum similarly noted how medical professionals were enlisted in the campaign to promote tear gas during the interwar years, as a way of constructing it as a tool of modern science rather than police weaponry. As Balmer, Spelling and McLeish showed, the Himsworth case extended this approach in framing tear gas as a drug, which resulted in a focus on toxicity thresholds and the concomitant need for medical expertise. Furthermore, the 'largely consensual situation of receiving a drug was linked to non-consensual crowd control. Arguably, this move also created an elision throughout the report between non-lethal and non-violent, because use of CS gas to quell civil disobedience still remains an act of violence'.[12]

Finally, it is also possible to broaden our frame of reference to rethink the separation of chemical and biological welfare. Etienne Aucouturier took such an approach by suggesting that the intended target of both CBW (as they are currently defined) is physiology or living systems, and arguing that the two might rather be unified under one concept of 'physiological warfare'. According to this line of thinking, distinguishing between chemical warfare and biological warfare (BW) only serves to further muddy the task of identifying civilian from military technology in verification measures, as well as defensive activities from offensive activities. Moreover, this approach also gives more visibility to the damage these weapons inflict upon ecological bodies and the environment.

REVISITING: ARTICULATIONS

Distinctions allow us to define, categorize, separate and group things in the world around us. However, the process of making distinctions is itself bound up with contexts, knowledges, politics, technologies and forms of expertise. Situated within these, chemical bodies take on different forms, with some features brought into focus and others fading away, gaining (il)legitimacy in certain contexts for particular roles but not others.

Aucouturier's notion of physiological warfare unified chemical weaponry with biological weaponry according to the two technologies' shared purpose of 'targeting physiological functions'. This articulation re-contextualizes these forms of warfare in a way that gives precedence to their effects on living systems, with implications for identifying, regulating and controlling the use of these weapons. Aucouturier points out that ecological warfare, for instance, would count as a form of physiological warfare because it entails targeting living functions. On the other hand, the two projects that Rappert,

Smith and Gould recount in their chapter – *Dis-eases of Secrecy* and *Poisoned Pasts* – encourage a plethora of simultaneous articulations of a CBW programme. Rather than settling on a singular narrative of Project Coast, their exhibition and monograph re-articulate diverse engagements with the past by encouraging their audiences to approach the history of Project Coast from their unique perspectives.

Michael Crowley's chapter, then, can be read as a call to open up the CWC to a more diverse range of articulations of legitimate chemical control, highlighting how the lack of guidance regarding 'law enforcement' and 'types and quantities' approaches in the CWC has resulted in uncertainty and contested interpretation of these terms and the failure to adequately account for issues like contexts of use, means of delivery, infringement of rights and who is manufacturing what kind of technologies and why. As such, the contested nature of what is an acceptable or unacceptable law enforcement weapon employing toxic chemicals, under the CWC, has potentially opened the door for states (and corporations and non-state actors) to develop, acquire and (mis)use an increasing number of these technologies – notably certain wide-area RCA means of delivery and ICA weapons – and to (mis)categorize them as appropriate for law enforcement, when clearly they are not from many perspectives.

Many of the other chapters, on the other hand, examined how articulations have functioned historically. For instance, CW use in the interwar period coincided with a range of dynamic geopolitical and military pressures, each providing further rationales for the value and necessity of chemical apparatus, and in turn further demarcating 'in-groups' populations from 'out-groups'. Both Revill and Favera's chapter and Feigenbaum's chapter spoke to this dynamic extensively. The former showcased a number of occasions on which military bodies turned to chemicals when they struggled to deal with uprisings, insurgencies and colonial battles with conventional weaponry. By contextualizing chemical agents alongside conventional weaponry, these actors thus constructed them as a solution of imperial control. Conceptually this mindset was grounded in the imperial notions of science as a form of 'civilizing' and progressing societies.

Yet, as Feigenbaum showed, at different points in time, British policy makers appealed to these articulations of chemical weapons to either legitimate or contest the use of CW in the colonies. With U.S. police adopting tear gas early in the 1920s, Britain still perceived police use of gas as undesirable, particularly in light of the recent gas use in World War I, as well as the Washington Treaty of 1922.[13] In the 1930s, tear gas began to find a place within imperial contexts, no longer broadly perceived as a war gas but rather a humane technology of science – not as part of a broader category that included war gases, but as part of an arsenal of bespoke police tools. This specific transformation was accompanied by a transformation of the police forces using tear gas – Feigenbaum points out how colonial policy makers initially considered their police forces ill-equipped to handle tear gas, but later referred to them as ordinary intelligent men who could handle the technology with little instruction

Elsewhere, recently Anna Feigenbaum has identified how 'all around the world people invent, adapt, and share techniques for resilience and resistance to teargas. In doing so, they care for each other. They transform this weapon into a collectivizing tool'.[14] From this perspective, these resistant actors are using their community and its knowledge of tear gas to re-assert their control over the state. Conversely, the proposed potential use of ICA weapons in counterterrorism or counter-insurgency operations might also be understood as part of broader narratives of national security, law enforcement and protection against non-state actors.

In the case of the Himsworth Committee, framing the issue as one of medical expertise and toxicity thresholds allowed CS gas to be given 'a clean bill of health' but not without tensions. If we think of Rappert, Smith and Gould's chapter as one that actively 'opened up' ways of articulating chemical histories, then Balmer, Spelling and McLeish's chapter conversely showed how a singular articulation of a chemical agent (CS gas) delimited investigation. The Himsworth Committee effectively assumed those involved in a protest to be legitimate targets in order to resolve the framing of CS as a drug. The value of a conception of 'chemical bodies' comes out prominently here – making judgements about chemical agents inextricably involves making judgements about bodies.

This is echoed in Mankoo's contextualization of tear gas as a form of civil defence protection. In a civil defence frame, tear gas was a way of imitating prospective enemy gas attacks, such that focus was no longer on whether or not its effects were acceptable, but on what potential they had to elicit responses from the public (such as donning gas masks). In doing so, the implicit assumption was made that tear gas was not harmful and that every member of the public was not just a legitimate target but also someone to be trained. This framing was further enforced by the fact that the public populations involved in the tests were made up of a significant proportion of women and children – the 'domestic' construction of their bodies legitimating the use of chemicals on them as a form of protection against wartime attack. This historical example can prompt us to consider how narratives of care and protection, for instance, feature within CW (or riot control) discourse today. What impact do they have, and whose interests do they represent? This is certainly of importance in relation to the contested discourse between proponents of the potential application of ICA weapons in certain counter-insurgency or counterterrorism operations, and the growing body of medical and scientific research highlighting the grave dangers to health and life of those who would be targeted by such weapons.[15] An even more critical step to take after this would be, then, to reflect upon what normative commitments lie within framing the justifiability of ICA weapons as something to be, in large part, decided by science or medicine: Who and what does this body of medical and scientific research represent? What and who does it foreground, and why, and what and who does it omit and obscure? What if, to take an example, we took the task of choosing what kind of cultures we want the military and police to have in our society as our starting point instead?

The choice to focus on socio-technical complexity through the notion of articulation as done in this volume also demands an openness to working with uncertainty, and disharmonious perspectives on technological trajectories, appropriate use of force and so on. When examining the history of Project Coast, for example, it is clear that searching for a singular account of the programme's activities would be unproductive, if not even irresponsible in certain respects. Instead, drawing out the role that chemical agents played in a range of accounts – each featuring different technologies, actors and expertise – reveals injustices, transgressions, harms and systems of control that will at some points be in tension with one another. This is tangibly demonstrated by noting the existence of the various 'epistemic constituencies' within which CS gas featured throughout the 'weapons as drugs' narrative of the Himsworth Committee. Making the decision to view CS through a 'medical framework' involved making some transgressions or actions less visible – as mentioned, this implied constructing everyone involved in a protest as a legitimate target, and as such, made chemical offences against potential 'bystanders' less visible. Attempts to bring Wouter Basson to account for his role in heading Project Coast, on the other hand, highlight how these frameworks (in this instance, one of legal expedience) can make certain chemical weapons use more visible. Because he was (and at the time of writing still is) a medical professional, his actions were deemed illegitimate because they lay outside the specific community bounds of acceptable professional medical conduct.

REVISITING: EXPERIMENTATION

This use of chemicals to sort and order bodies and spaces also constitutes forms of experimentation, some more explicit, others more subtle. Revill and Favera pointed out how chemical weapons use in the interwar period was often initially conducted on experimental grounds before becoming situated in more explicitly aggressive militaristic activities. Chemical experimentation on bodies often takes place in spaces where distinctions between what is permitted and necessary, and whose lives count, are being actively negotiated and contested. This was noted by Guillemin with regard to Japan's experimentation with captive prisoners as well as Rappert, Smith and Gould with regard to Project Coast's experimentation on animals. Similarly, the function of civil defence gas tests as a form of human experimentation was enabled by the fact that usual distinctions of 'domestic' and 'military' were in a state of wartime suspension. The legitimacy of the use of chemicals to protect, and experiment with, the bodies of the British public was heavily entangled with the geopolitical through Britain's wartime agenda.

Many of the chapters examined instances of more figurative forms of experimentation. Crowley highlighted how both states and weapons manufacturers are engaging in a kind of reckless experimentation with the boundaries of what is and is not allowed by the CWC, and by proxy, what kind of experimentation with what kind of chemical harms against the body are acceptable in both military and civilian situations. We might also ask whether, by classifying CS as a drug, the Himsworth Committee constructed prospective use of CS gas on protestors as a form of medical experimentation. Certainly, this classification did entail a form of experimental simulation – one that used 'observations from the Northern Irish and other geo-political events and [combined] them with experimental data to create predictions of the effects of CS in any situation'.[16] Furthermore, the evidence used to determine threshold levels of CS exposure was itself drawing from experiments on animals as well as reports from Northern Ireland (and elsewhere).

Conversely, Anna Feigenbaum's work here also detailed what led to the experimental British search for a new tear gas agent after World War II. In the Peshawar prison altercation, one man died and many more were injured during the use of British-supplied tear gas, though police intended it to be non-lethal. The messiness of this particular situation – that the police were unable to control the effects of gas as desired – halted the British manufacture of No. 92 tear gas grenades and led to experimentation in search for a new alternative. The search, which ultimately resulted in the British adoption of CS gas years later, exemplifies how experimentation can also be the result of perceived transgressions and messy scenarios that lack order.

As evident from the case of Northern Ireland and also other contexts under scrutiny by the contributors, 'experimentation' was associated with acts of control but also demarcated the limits of control. As one consideration, the need for experimentation is often a reaction to what is perceived by some as a disorderly situation. As another consideration, those individuals and places meant to play the part of subjects in experiments often refuse the role being ascribed to them. In attempting to impose control, experiments often signal the inability of authorities to dictate responses. In these senses, 'mess' and 'order' came bundled together.

INJUSTICES, ACCOUNTABILITY AND TRANSGRESSIONS

All of the chapters in *Chemical Bodies* have spoken to how the geopolitical is bound up with the use of chemicals to control the body – how it directs and has been directed by the distinctions that are used to assess and define chemical weapons, how it pervades the contexts, technologies and types of expertise in which they are situated as well as how it has directed and is directed by attempts to experiment with bodies and boundaries. As we have seen from the case of the IMTFE, Project Coast, and Britain's World War II gas tests, many of the analyses here shed light on what has been absent from both chemical weapons histories and policies, what has been missing as well as what has been present.

Why is this important for those working in fields related to chemical weapons and technologies of control? What broad implications does this approach have for these fields? For the remainder of this closing chapter, we wish to contemplate the value of this way of thinking about chemical weapons. Whose experiences come to light? Can it inform how we think about injustices or 'transgressive' technologies? What asymmetries does it reveal – asymmetries of power, knowledge and voice? What types of futures does it reveal?

In chapter 1, 'Transgressive Chemicals', we contended that 'distinctions matter. They cordon off parts of the world from each other'. As such, crossing-established distinctions often constitute injustice, or inappropriate, unacceptable action of some kind. However, as many chapters in this volume have shown, such transgressions can and do occur insidiously, particularly when the use of chemicals takes place in contexts that call into question obscure and suspend boundaries. Herein, actors (attempt to) order the world in new ways, making certain technologies, actors and forms of knowledge more or less visible.

These points resonate with ones made elsewhere, particularly literature on the production of knowledge and ignorance. In her analysis of Belarusian public knowledge of radiation health effects in the aftermath of the Chernobyl disaster, Olga Kuchinskaya highlights how the Belarusian government provided free lunches as a form of compensation to children who showed high levels of radiation exposure.[17] She suggests that school radiation tests were a way of making the health of effects of radiation (in)visible. This policy effectively attempted to address a health problem with a form of mundane economic compensation (free school lunches), but Kuchinskaya identifies how, for particular Belarusian publics, it instead meant that the radiation in their area was of no concern, given the minimal and mundane form of economic compensation. Health effects associated with radiation were interpreted and understood through this economic framing, and this made certain actors, technologies and knowledges more visible (and others less visible). Similarly, articulations of chemical technologies and their effects involve the re-ordering of what is visible (in the case of Chernobyl, this re-ordered what was visible to those living with the radiation itself).

These kinds of transformations, reminiscent of Balmer, Spelling and McLeish's notion of 'epistemic constituencies', have been prominent throughout these chapters. These transformations can and should raise concerns. It is unsettling that framing CS as a medical and toxicological issue changed the focus of the Himsworth investigation, concealing certain details while bringing others to the fore. The negotiation of visibility is central to establishing injustices and transgressions. Given that chemical and biological weapons research predominantly takes places in conditions of secrecy, much is at stake in the negotiation of visibility and invisibility. In some instances, for instance, the limits of what is known might effectively cut off scrutiny. In other instances, what is known suggestively signals the overall outline shape of what is not.

This is not to say CBW programmes are too complex, secret or opaque for us to make any judgements about transgressive or unjust behaviour altogether. This volume has explored more than a number of examples of chemical weapons use and research programmes being held to account. Elsewhere Ulf Schmidt has documented in *Secret Science* the British legal inquest in which programmes of human experimentation at Porton Down were held to account by the ex-servicemen who were experimented upon.[18]

Rather, we are focusing on limits of what is known as a means to critically examine our notions of 'transgression'. What does it mean for technology to be transgressive? Is it enough to simply ask what it is that chemical weapons technologies are transgressing across? To take transgression as simply an act of crossing established boundaries? Our approach here suggests that it is not. If we return to thinking of co-constituted chemical bodies, and to the idea of ontological multiplicity, questions about whether a boundary has been transgressed must also be thought of as questions regarding the legitimacy of the boundary itself. Articulating transgression is itself an act of re-ordering the bounds of what is acceptable, such that pointing to a transgression (an abuse of power) is at once a claim to power in and of itself. Drawing from examples from within this volume, the following list explores some of the boundaries that chemical technologies have been positioned as transgressing (at least by some), and opens up what normative commitments might be taken for granted within these articulations of transgression:

- **Norms**: Transgression involves the crossing of boundaries. Using a chemical weapon as a means of warfare would cross the established distinctions regarding what is within the bounds of acceptable means of warfare. This could take place with respect to various dimensions:

    - Moral norms: These could be more tangible, explicitly stated norms or implicit, subtle ones – commentators have noted the moral opprobrium surrounding CW that has been enacted throughout the twentieth century, and the role taboo has played in shaping their trajectory.[19] Chemical weapons have been imbued in discourse with an inherently morally aberrant status that itself becomes a strategic tool with which actors can situate transgressions. In these instances, it is commitments to the ideas of morality that are taken for granted.

- o Transgressing institutional, cultural and legal norms: These are linked to moral norms, and shape one another, but they are not one and the same, and they may not always align. In the case of IMTFE, for example, CW use transgressed across moral norms but was not taken as transgressing across the legal norms of the GP. Crowley here and elsewhere has explored the wide range of potentially applicable legal and institutional norms that could be brought to bear to prohibit, constrain or address development or (mis)use of certain weapons employing toxic chemicals. Although these certainly include relevant arms control and disarmament treaties (i.e. CWC and BTWC), they also include international human rights law, international humanitarian law and international criminal law as well as certain UN Drug Conventions. In addition, institutional and professional ethical guidelines might prevent the involvement of medical professionals in the development and use of at least some of these weapons.[20] Norms such as these, though, also operate both as statements about the world and how it should be governed. Defining something as an institutional issue, an issue for international law or a medical issue, all cut off certain parts of the world while bringing to light others. Setting these boundaries is also an act of deciding when and where who gets to speak for whom about what.

- **Bodies**: Chemical technologies breach bodies. They can permeate and convert our outer skin in ways held with revulsion. While some bodies are targets, effects upon others might be unintended. In an instance of tear gas use upon a rioting crowd, effect on bystanders in the vicinity can be held as lamentable. Herein though the applicability of the terms hinges upon distinctions of who counts as a legitimate target, distinctions that are dependent on articulations. The Himsworth Committee's 'CS as a drug' approach, for instance, renders effects on potential bystanders as legitimate. Alternatively, the use of CW in human experimentation, such as that which took place during Project Coast, as well as that which both Ulf Schmidt and Rob Evans have highlighted, also involves wrongdoing against bodies.[21] With the use of chemicals for experimentation, transgression is related to notions of informed consent, autonomy and the rights of those individuals who are the test subjects. Nonetheless, articulations still shape what counts as transgression in these contexts. The need for civil defence in wartime meant that indiscriminate use of tear gas to train the public was relatively socially acceptable in World War II Britain, whether or not people had consented, for instance. Establishing whether chemicals have transgressed bodies, then, does not account for the fact that some bodies come to matter more than others, or that these boundaries become legitimate to cross in some contexts but not others.[22] The body becomes a staging ground for the politics of transgression. Furthermore, as chemicals permeate space as well as the body, they limit what the body can do by making parts of the world less livable.[23] In this sense, they call into the question the boundaries of bodies themselves, as some bodies can traverse spaces that others cannot – those bodies with a gas mask in a tear gassed crowd, for instance, become more livable than those without. Scholars have long challenged the notion that the body and its limits can be read as anything fixed.[24] Rather, then, we must 'treat the creation of lines of separation and difference as always contingent, often strategic' to use Lucy Suchman's phrasing regarding body-machine distinctions (and in our aforementioned example, we might think of the gas mask as the 'machine' here).[25]

- **Categories**: Returning to the theme of distinctions, chemical technologies also cross the various conceptual categories that attempt to define them. For example, a tear gas canister can cause fatal burns (as in the case explored by Feigenbaum), or can kill on impact, such that the weapons system moves outside of the 'non-lethal' category. Within some contexts (e.g. enclosed spaces, as noted by Crowley), the use of 'non-lethal' tear gas can itself be fatal. As we have seen throughout the volume, this transgression is limited to not only the boundaries of lethal/non-lethal but also boundaries such as law enforcement/armed conflict, experiment/field, chemical/biological/physiological and national/international. What is so noteworthy about chemical agents is how starkly and dramatically their categorical status can change. What was once held as a means of saving life can switch to being an instrument of repression, what was once designed for peaceful civilian use can be regarded as enabling malign ends. Yet these categories are themselves claims to power – they take for granted how the world should be divided up. Thus, determining whether a chemical agent is transgressing these categories involves consideration of the legitimacy of the categories themselves. Deciding whether a chemical is non-lethal comes hand in hand with deciding what the bounds of lethality and non-lethality are. And these bounds determine how the world should be governed – in the words of Geoffrey Bowker and Susan Leigh Star, the 'thicket of classifications is both operative (defining the possibilities for action) and descriptive'.[26]

In these ways, the articulation of transgression is a way of foregrounding (or obscuring) certain forms of infringement – sometimes at the expense of others. If transgression entails an asymmetry of power, in that it involves the act of crossing the boundaries of what is acceptable – by certain actors often at the expense of other actors – then articulating transgression is also an act of power. That is to say, certain actors can define the boundaries being crossed at the expense of other actors. With this in mind, it is unsurprising that questionable deeds sometimes only come to light from particular perspectives – often those that might be closest to those of the victims – and only through considerable work. This is perhaps best exemplified by the *Poisoned Pasts* exhibition on Project Coast. Some transgressions only exist within certain articulations of chemical bodies. Contesting an articulation where a transgression does not exist in favour of one where it does is therefore an act of 'making' a transgression real, of enacting it and bringing it into being. If a transgression does not exist within dominant articulations (say, long-standing institutional frameworks, or cohere with state level socio-technical imaginaries, to use a science and technology studies (STS) term),[27] then the 'work' for actors who contest these articulations to make this transgression visible is that much more.[28]

Yet, there are also times when particular instances of use become more broadly consensual points of focus. The author of the *Denver Post* commentary, for example, is questioning why killing with chemicals is a point of focus in the international arena, when killing with conventional weaponry is occurring in the same place to a much larger extent. An implication of this critique is that conventional weapons use should be held to account according to similar standards as chemical use. However, as the chapters in this volume have shown, notions of chemical weapons are entangled with particular contexts, expertise, technologies, institutions and so on. While calling for condemnation of deaths from conventional weapons use in Syria alongside CW use is a commendable objective, the fact that conventional weaponry is tied up with a multitude of other actors and spaces begs the question of this condemnation implied for conventional weaponry.

It is true that the exceptionality chemical warfare holds elevates its visibility in certain instances, such as the case in Syria, particularly within visual media and news coverage of Western nations. The commentary rightly asks why this is the case – but by honing in on this distinction, it potentially misses the more pertinent question: Why has this particular use of CW come to the fore now, and here, within this conflict? To take a case in Syria, the use of CW at Ghouta in 2013 received a significant amount of attention from both the media and states parties, but allegations of CW prior to this received relatively little. A paper summarizing a 2016 Workshop on Syria by the Harvard Sussex Program noted that this could have in part (particularly from a media perspective) been due to reticence towards making such accusations in light of the conjectures regarding weapons of mass destructions (WMDs) in Iraq during the previous decade.[29] This mitigated enthusiasm for military action and enabled 'certain actors to exploit uncertainty'.[30] Furthermore, the pre-Ghouta allegations were a major source of division for states parties in the Third Review Conference of the CWC, tying up geopolitical stances with making formal judgements on chemical weapons use, thus rendering attending to chemical warfare more complex.

The Syria Workshop paper also highlights how attempts to control and condemn chemical weapons use do not necessarily result in providing victims with more voice, much like we have seen in cases such as the IMTFE, Crowley's analysis of the CWC[31] and Wouter Basson's criminal trial. In fact, 'Although the destruction of chemical weapons [in Syria] was perceived as a huge success for the disarmament community, few Syrians spoke of this process in positive terms. Rather, the predominant perception of many Syrians was that this process legitimised the Assad regime and exacerbated the levels of brutality without effectively stopping the use of chemical weapons'.[32] Measures of control in a geopolitical space, while valuable in their own right, do not translate to a reduction in the sense of vulnerability that those living within spaces of conflict might feel with regard to their bodies.

This volume has not attempted to suggest a singular way to approach CW debates or control. Rather, we are advocating the value of a multifaceted approach that involves examining the social and technological relations involved in the chemical apparatus, which can reveal a range of perspectives that shift in and out of harmony with one another. We remain aware that taking any form of action on the basis of such analysis – whether policy, activist action or other – involves sorting through and making decisions about which perspective and relations are most important in a given situation. Moreover, a truly reflexive approach would also take into account the processes that shape how we go about this multifaceted analysis in the first place, and how our work in and of itself sorts and orders bodies.

However, this issue by no means renders our approach useless. Rather, surveying chemical bodies from a range of viewpoints can help to inform scholarship, activism and policy in some of the following, more nuanced, ways:

- **Highlighting Divergent/Convergent Articulations**: In that attempts to (de)legitimate chemical weapons are ways of reconfiguring networks of contexts, technologies, expertise (etc.), analysis of these positions will reveal areas where different attempts to do so diverge and converge. These may in some cases reveal surprising harmonies or disharmonies – for example, that the destruction of chemical weapons does not translate to success in the minds of many Syrians; or that wartime use of tear gas for public tests was actively encouraged by some of those being gassed. However, the rationales behind converging articulations might be radically different. When that is the case, these can be identified, evaluated and (in some instances) held to account.
- **Suggesting Where to Look**: Many chapters in this volume have discussed how what is known about chemical bodies is contingent upon what is unknown, missing or obscured (and vice versa). Questions that follow related to this point are as follows: Where do uncertainties, ambiguities and unknowns lie, and how do different viewpoints deal with them? Who

decides how to distinguish between what is relevant and what is not? Are there ways of opening this up? (The *Poisoned Pasts* exhibition certainly operated as one way of doing so. Etienne Aucouturier's suggestion of the notion of physiological warfare is another way of doing so, which brings issues such as ecological warfare closer to centre frame.)

- **Identifying Chemical Futures**: Finally, thinking in terms of chemical bodies has potential to open up the way we think about futures. This includes not only the technological trajectory of chemical technologies but also the imaginaries of the distinctions and articulations – the kinds of societies, contexts and institutions – that these feature in. For instance, focusing on enforcing the declaration and destruction of chemical weapons conjures a future that is significantly orientated around preventing their offensive application but does not necessarily entail a future in which dual-use technologies, or chemicals with a multitude of applications, are being continually monitored, evaluated and controlled. Michael Crowley's chapter was in many ways a call for us to be more aware of the range of futures – some more messy and unruly than others – that are, right now, open to ICA weapons, RCAs and the growing range of associated delivery mechanisms. Shifting the focus from a singular trajectory (the destruction of Cold War-era state CW programmes) to the possibilities posed by broader security challenges is key for policy makers who wish to progress chemical control in an anticipatory and democratic manner. The category of 'law enforcement purposes' is not clearly defined and remains open to a range of interpretations. This is not in and of itself negative, but these interpretations (and the implications they have for a range of groups) must be assessed alongside one another if the future of chemical control is to be responsible.

NOTES

1 The notion of a co-constitutive relationship here draws from ideas within co-productionist work in the field of STS. This approach argues that thinking of natural and social orders as being produced in tandem provides those studying history, political and cultural formations, with explanatory power. See Sheila Jasanoff, ed., *States of Knowledge: The Co-Production of Science and Social Order* (London: Routledge, 2004).

2 Annemarie Mol, *The Body Multiple: Ontology in Medical Practice* (Durham: Duke University Press, 2002), 6.

3 Justin McBrayer, 'Why Should the Line Be Drawn at Chemical Weapons?' *Denver Post*, 10 April 2017, accessed on 2 December 2017, available at http://www.denverpost.com/2017/04/10/why-should-the-line-be-drawn-at-chemical-weapons/.

4 J. F. C. Fuller, *The Reformation of War* (London: Hutchinson & Co., 1923); J. B. S. Haldane, *Callinicus: A Defence of Chemical Warfare* (London: Kegan Paul, Trench and Trubner & Co., 1925).
5 Amnesty International News, 'Spain, Accountability Urged for "Appalling" Migrant Deaths in Ceuta', 14 February 2014, available at https://www.amnesty.org/en/latest/news/2014/02/spain-accountability-urged-appalling-migrant-deaths-ceuta/; Amnesty International News, 'Spain: Ceuta Migrant Tragedy – Deplorable Disregard for Human Life', 6 February 2015, accessed 4 December 2017, available at https://www.amnesty.org/en/latest/news/2015/02/spain-ceuta-migrant-tragedy-deplorable-disregard-human-life/; P. Kingsley, 'How Did 37 Prisoners Come to Die at Cairo Prison Abu Zaabal?', *The Guardian*, 22 February 2014.
6 See Jasanoff, *States of Knowledge*.
7 Army Field Manual, *Volume 1: Combined Arms Operations. Part 10: Counter Insurgency Operations (Strategic and Operational Guidelines)*, Revised and Updated Version, Issue 2.0 (March 2007), B-7-B-1 and B-7-B-2, available at http://www.operationtelic.co.uk/documents/uk-coin-operations-2007.pdf.
8 Ibid., B-7–6.
9 Ibid.
10 Ibid.
11 For further discussion on these issues, see Michael Crowley, *Chemical Control: Regulation of Incapacitating Chemical Agent Weapons, Riot Control Agents and Their Means of Delivery* (Basingstoke: Palgrave Macmillan, 2016).
12 See Balmer, Spelling and McLeish's chapter in this volume.
13 Simeon Shoul, 'British Tear Gas Doctrine between the World Wars', *War in History*, 15, no. 2 (1 April 2008), 168–190.
14 Anna Feigenbaum, *Tear Gas: From the Battlefields of WWI to the Streets of Today* (London; Brooklyn, NY: Verso Books, 2017), Chapter 8.
15 Kai Ilchmann and James Revill, 'Chemical and Biological Weapons in the "New Wars"', *Science and Engineering Ethics*, 20, no. 3 (September 2014), 753–767; Crowley, *Chemical Control*, Chapter 2.
16 See Balmer, Spelling and McLeish's chapter in this volume.
17 Olga Kuchinskaya, *The Politics of Invisibility: Public Knowledge about Radiation Health Effects after Chernobyl* (Cambridge, MA: MIT Press, 2014).
18 Ulf Schmidt, *Secret Science: A Century of Poison Warfare and Human Experiments* (Oxford: Oxford University Press, 2015).
19 Richard MacKay Price, *The Chemical Weapons Taboo* (Cornell University Press, 1997).
20 Crowley, *Chemical Control*, Chapters 5–13.
21 Schmidt, *Secret Science*; Robert Evans, *Gassed: Behind the Scenes at Porton Down* (London: House of Stratus, 2000).
22 Julie Guthman and Sandy Brown, 'Whose Life Counts: Biopolitics and the "Bright Line" of Chloropicrin Mitigation in California's Strawberry Industry', *Science, Technology, & Human Values*, 41, no. 3 (2016), 461–482. Guthman and Brown's work is a compelling piece of recent scholarship that shows

how some bodies come to count more than others in the context of fumigant chemical use.

23 Peter Sloterdijk, 'Airquakes', *Environment and Planning D: Society and Space*, 27, no. 1 (2009), 41–57. Sloterdijk argues that the twentieth century saw an emergence of what he calls 'atmosterrorism': assault not on the bodies of enemies but on their environment.

24 Donna Haraway, 'A Cyborg Manifesto: Science, Technology, and Socialist-Feminism in the Late Twentieth Century', in *Simians, Cyborgs and Women: The Reinvention of Nature* (New York: Routledge, 1991), 149–181; Mol, *The Body Multiple*; Elaine Scarry, *The Body in Pain: The Making and Unmaking of the World* (New York, Oxford: Oxford University Press, 1985); Lucy Suchman, *Human-Machine Reconfigurations: Plans and Situated Actions* (Cambridge: Cambridge University Press, 2007).

25 Lucy Suchman, 'Situational Awareness: Deadly Bioconvergence at the Boundaries of Bodies and Machines', *MediaTropes*, 5, no. 1 (2015), 1–24.

26 Geoffrey C. Bowker and Susan Leigh Star, *Sorting Things Out: Classification and Its Consequences* (Cambridge, MA: MIT Press, 1999), 326.

27 See Sheila Jasanoff and Sang-Hyun Kim, *Dreamscapes of Modernity: Sociotechnical Imaginaries and the Fabrication of Power* (University of Chicago Press, 2015).

28 Kuchinskaya (*The Politics of Invisibility*, 21) refers to the 'work' of articulation: 'The imperceptibility of radiation with the unaided senses thus means that radiation danger and its possible health effects might not be spontaneously obvious to those who experience them. The signs of radiation danger and its connection to actual or potential health effects have to be articulated: identified, explicated, and established as such. Radiation is not visible or observable to laypeople living in the contaminated areas without this work of articulation'. Our proposal here, however, combines this with Mol's ideas to suggest that the work of articulating something like a transgression does not simply involve making it visible but also making it 'real'.

29 James Revill, Caitríona McLeish, Steve Johnson, Alex Ghionis and Brett Edwards, 'Workshop Summary', Harvard Sussex Program Occasional Paper, Syria Collection (June 2016), 11–12.

30 Ibid., 12.

31 Crowley, *Chemical Control*, chapter 5.

32 Revill et al., *op cit.*

About the Contributors

Etienne Aucouturier is a philosopher and historian of science. His area of specialization is the history and philosophy of biology and medicine. He was previously head of Maison Française at University of Ghana and French Foreign Office, and obtained his PhD from Université Paris 1 Panthéon-Sorbonne in 2011. His research focuses on the misuses of biology and medicine, in particular on biological warfare programmes. He is currently based in Paris, where he also teaches philosophy.

Brian Balmer is professor of science policy studies in the Department of Science and Technology Studies, University College London. His research combines historical and sociological approaches to understanding the nature of expertise, particularly in the life sciences. He has particular interest in the history of biological and chemical warfare, the history of the 'brain drain' and the role of experimental volunteers in biomedical research. He has published widely on these subjects and is author of *Britain and Biological Warfare: Expertise and Science Policy, 1935–1965* and *Secrecy and Science: A Historical Sociology of Biological and Chemical Warfare*.

Michael Crowley is an honorary visiting senior research fellow at Bradford University and is also research associate with the Omega Research Foundation. He has worked for twenty-five years on arms control, security and human rights issues, including as executive director of the Verification Research, Training and information Centre (VERTIC), and has acted as chairperson of the Bio-weapons Prevention Project. He has managed the Arms Trade Treaty project at the Arias Foundation in Costa Rica and worked as senior arms trade analyst at the British American Security Information

Council (BASIC). He has also had several research and policy positions with Amnesty International, both in the UK Section and at the International Secretariat. He holds a doctorate from Bradford University on the regulation of riot control agents and incapacitating chemical agent weapons.

Marcos Favero has an undergraduate degree in international relations from the São Paulo State University and a master's degree in peace, conflict and development from the University of Bradford. He works as an international consultant at ProDialogo, a researcher for foreign policy, as well as a fellow facilitator at Soliya. His research focuses on dialogue and intractable conflicts.

Anna Feigenbaum with her work focuses on communication and social justice. She is a co-author of *Protest Camps* (Zed 2013) and the author of *Tear Gas: From WWI to the Streets of Today* (Verso 2017). In addition to appearing in academic journals, her research can be found in the *Guardian*, the *Atlantic*, Al Jazeera America, New Internationalist, Open Democracy and Waging Nonviolence, among other outlets. Anna runs the Civic Media Hub Datalabs project at Bournemouth University, delivering workshops and teaching on data storytelling. Her consultancy work includes Amnesty International and Public Health Dorset.

Chandré Gould is a senior research fellow in the Justice and Violence Prevention Programme at the Institute for Security Studies in South Africa. Her most recent projects, exploring the Apartheid-era chemical and biological warfare programme in South Africa, have included collaborations with Brian Rappert and artist Kathryn Smith. These are an exhibition titled *Poisoned Pasts*, curated by Smith, Rappert and Gould, and a book titled *Dis-eases of Secrecy: Tracing History, Memory and Justice* (Rappert and Gould). Both projects have sought to challenge and disrupt the expectations and aspirations associated with revealing hidden pasts.

Jeanne Guillemin is a senior advisor at the MIT Security Studies Programme within the Center for International Studies. Trained at Harvard and Brandeis Universities and the Sorbonne and a former professor of sociology at Boston College, her past publications include numerous articles on the history of biological and chemical weapons and four well-known books, one on the 1979 outbreak of anthrax in Sverdlovsk (Union of Soviet Socialist Republics), second on state biological weapons programmes, third on the 2001 anthrax letter attacks in the United States and, most recently, American Atrocities, about the U.S. military intelligence's secret intervention at the Tokyo Trial to suppress evidence of Japanese germ and chemical warfare. In addition to consulting for government agencies, she has served on several national commissions and medical ethics boards.

Alex Mankoo is a PhD candidate in the Science and Technology Studies Department at University College London and a visiting fellow in the Science, Technology and Society Program at the Harvard Kennedy School of Government (Spring 2018). His research interests include the history of chemical and biological weapons, intersections between science and security and the sociology of security technologies more generally. His PhD research, funded by the Economic and Social Research Council (ESRC), focuses on the technological trajectory of tear gas in mid-twentieth-century Britain and the types of legitimacy it gained during this period.

Caitríona McLeish is a senior fellow at SPRU at the University of Sussex and director at the Sussex end of the Harvard Sussex Program on chemical and biological weapons. The main focus of her research is on the governance of dual use technologies in regard to the effective implementation of the prohibitions under both the Chemical Weapons Convention and Biological Weapons Convention and how to create effective mechanisms to prevent the misuse of legitimate science and technology. Her recent work includes analysis of past offensive and defensive chemical and biological warfare programmes; chemical weapons use and alleged use in Syria and Iraq; assessments of the impact that contemporary dual use policies are having on the innovative capacity of scientific and industrial communities; and examination of the roles of industry and civil society in chemical and biological disarmament efforts. She was also co-investigator on an AHRC-funded project examining the history of the context for the negotiations of the Biological Weapons Convention.

Brian Rappert is a professor of science, technology and public affairs at the University of Exeter. His long-term interest has been examination of the strategic management of information; particularly in relation to armed conflict. His books include *Controlling the Weapons of War: Politics, Persuasion, and the Prohibition of Inhumanity*, *Biotechnology, Security and the Search for Limits* and *Education and Ethics in the Life Science*. More recently, he has been interested in the social, ethical and political issues associated with researching and writing about secrets, as in his books *Experimental Secrets* (2009), *How to Look Good in a War* (2012) and *Dis-eases of Secrecy* (2017). For more information, see http://brianrappert.net/.

James Revill is a research fellow with the Harvard Sussex Program at SPRU, University of Sussex. Revill's research interests focus on the evolution of the biological and chemical weapons treaty regimes, especially analysis of the

Biological Weapons Conventions, the interplay between science and security and the role and risk of non-state groups in chemical, biological and explosive terrorism. He has published widely in these areas, including a book-length study on the history of Improvised Explosive Devices (IEDs), several papers on issues pertaining to the BWC and more recently an assessment of the risk of chemical and biological terrorism.

Kathryn Smith is an interdisciplinary visual and forensic artist, and curator working across art and science. Operating at the interface of the studio/lab, the archive/library and the museum/gallery, her practice evidences a critical poetic relationship with the forensic imaginary and related processes of truth making, mainly through photographic, video and other documentary practices. She has been producing and exhibiting multimedia installations, site-responsive actions and art writing internationally since 2000, held by the notional project space, *serialworks*. Her interest in co-authorship and interdisciplinary collaboration as research methods is reflected in several major curatorial projects, including *Poisoned Pasts* (2016). Currently a postgraduate researcher in Face Lab, Liverpool John Moores University, she also holds a senior lectureship in fine art at Stellenbosch University. A selection of her projects and publications can be accessed via www.kathrynsmith.info.

Alex Spelling spent 2013–2016 as a research associate in University College of London's (UCL) Science and Technology Studies (STS) Department on an AHRC-funded project, 'Understanding Biological Disarmament: The Historical Context of the Origins of the Biological Weapons Convention'. From 2008 to 2013, he was a teaching fellow in American History at the Universities of Southampton and Aberdeen, having been awarded a PhD from the University of Nottingham for a study looking at Anglo-American relations during the Nixon administration. He is currently an honorary research associate in STS Department at UCL. His research interests are in Cold War history from a transatlantic perspective, in particular the U.S.–UK relationship and arms control issues relating to chemical and biological weapons.

Index

Lightning Source UK Ltd.
Milton Keynes UK
UKHW040932250220
359280UK00003B/389

9 781786 605863